Seeking Ultimates

'[Landsberg]...offers us his vision of a great spectrum of topics, ranging from fundamental particles to models of the universe, from the periodic table to the origins of life, from the global energy supply to Gödel's theorem. We go on a...ride starting with thermodynamics (...mass, perpetual motion), moving on to...elements, particles, forces...continuing to time and entropy...(self-organization,...chaos and...the origins of life), and quantum theory (waves and particles, wave functions and probabilities, quantum gravity, nonlocality, Schrödinger's cat...), arriving finally at cosmology (...black holes, ...physical constants, the anthropic principle), mathematics (...complexity), and even religion. Landsberg treats all of this and more in his inimitable style: terse, concise, and to the point, but chock full of insights and humor.'

'This book is not only illuminating but also entertaining. It is embellished throughout by illustrations,...examples of correspondence between scientists, and anecdotes.... Each chapter is given a hero...Pascal, Rumford, Mendeleev, Boltzmann, Darwin, Planck, Einstein, Eddington. These serve...to show how important a love of science for its own sake is to genuine progress in understanding.'

'I heartily recommend this book.... If you have not been waiting for this book, you should have been, and if you have not read it yet, you should.'

American Journal of Physics
October 2000

Seeking Ultimates
An Intuitive Guide to Physics

Peter T Landsberg

University of Southampton

Routledge
Taylor & Francis Group
LONDON AND NEW YORK

First published 2000 by Institute of Physics Publishing

2 Park Square, Milton Park, Abingdon, Oxfordshire OX14 4RN
52 Vanderbilt Avenue, New York, NY 10017

Routledge is an imprint of the Taylor & Francis Group, an informa business

First issued in hardback 2019

British Library Cataloguing-in-Publication Data

A catalogue record for this book is available from the British Library.

Library of Congress Cataloging-in-Publication Data are available

Cover Design: Kevin Lowry

ISBN 13: 978-0-7503-0657-7 (pbk)
ISBN 13: 978-1-138-42976-5 (hbk)

Contents

Introduction

An intellectual discipline is one thing—a book about it is another. Take poetry, for example; you can write it down on a piece of paper or read it in a book. But you *live* poetry by knowing it in your heart and mind, by reciting it, by lending emphasis here and a pause there. Similarly with a language—you can learn French from a French grammar or from a book of French songs. But you live it by speaking it, even by acting it. The shoulders might move for '*je m'en fou*' and you may nod your head '*Voila!*' So the discipline itself is different from its version as written down in a book. The book can enable you to 'live' it, by putting something of yourself into it.

Similarly, when lecturing on mathematical physics, you might tell students to 'forget' the mathematics, once they have understood it, and to try to appreciate what has been achieved in an intuitive manner—to absorb it into their bones, as it were—using physical insight. This is usually found to be a hard task, but an important one, and gets close to what I have called 'intuitive' in the title of this book.

In this way we arrive at 'popular science'. This is an important activity, for when the average person contemplates this universe, and the science which governs it, he must be excused for feeling rather confused by the language and by the details. Biology, psychology and even the brain are also at least partially physics-based; many of the concepts used are remote from normal experience, and the arguments can be mathematical. This book may be a help.

Only a few experiments are discussed in this book, although they provide the main mechanism for advancing science. We *do* science, for example, by heating a wire in a flame and seeing it turn blue; or by timing the oscillations of a suspended spring; or by studying the flight of a ball. Then we may develop equations to describe the trajectory of the ball. But that has no place in this book. Thus our constraints are rather severe. But we still want to attain an appreciation of the results and arguments of science in order to obtain an intuitive grasp of the connections between various phenomena; say, between light and gravity. This can be done, as shown here, but it requires some work on the part of the reader: at the very least he or she will have to turn pages forward and backward in order to understand the concepts, even though they may be standard ones (examples might be 'photons', 'antimatter', black holes', etc). Our constraints (few experiments, no mathematics) thus match those for books on poetry and French (no singing, no acting, no reciting!).

People have written about 'the end of science' and a 'theory of everything', and it has been said that with science as it is there may be 'no room for a creator'. The average person's gut reaction that such notions cannot be strictly correct is here vindicated as part of the text. That does not mean that we have no excitement. Some very unexpected effects are noted in the course of the discussion, and there is also some fun to be had.

I show where there are gaps which are being filled, but also that there are gaps which are more lasting features of the world as we see it. Discussions of entropy and time, the chemical elements and elementary particles, chaos and life, form part of this story, which starts with simpler ideas such as temperature. Later we explore quantum theory and cosmology. In all cases we look for 'ultimates'. Thus, we speak of 'isolated systems'—do they actually exist? Or does Newtonian mechanics really always predict exact results? Incompletenesses and uncertainties in both physics and in mathematics have to be faced, leading eventually to a discussion of God and human happiness in the light of what has been found.

Acknowledgments

I want to thank my family for their help and support and the University of Southampton for facilities made available to me. In particular, I wish to thank my 'IT consultants' Jeff Dewynne, Alistair Fitt and Colin Please.

Various colleagues read parts of the manuscript and I thank them for their comments. They are: Dr V Badescu (Bucharest), Dennis Blumenfeld (Chicago), Sir Hermann Bondi (Cambridge), Dmitry Bosky (London), Lajos Diosi (Budapest), Freeman Dyson (Princeton), Brian Griffiths (Southampton), Gareth Jones (Southampton), Andrew Kinghorn (Southampton), Max and Olivia Landsberg (London), John Liakos (Northampton), Robert Mann (Waterloo), George Matsas (São Paulo), Gunther Stent (Berkeley), Manuel Velarde (Madrid), James Vickers (Southampton) and notably Garry McEwen (Southampton), whose construction of, and help with, table 3.2 was particularly helpful.

For comments on Chapter 6 I want to thank Avshalom Elitzur (Jerusalem), Asher Peres (Haifa), Abner Shimony (Boston) and Andrew Whitaker (Belfast).

For comments on Chapter 7 I want to thank Tony Dean (Southampton), Jeremy Goodman (Princeton) and Malcolm Longair (Cambridge).

Chapter 1

What this book is about

1.1 Introduction

Our wish to understand the cosmos takes us to physics! It is the most fundamental of the sciences: even in biology or studies of the brain, the concepts from physics are essential. The snag is that physics has the reputation of being mathematical and hard to understand. We get around this problem here by the use of intuition. That is my first purpose. There is no mathematics in this book.

A red thread runs through this work to show that things are not as cut and dried as people often think: I emphasize, and that is my second purpose, that the notion of incompleteness is central to the whole of science.

1.2 My story

It may help if I tell you first a little about myself. In the troubled atmosphere of 1939, when I had just arrived in England and I had to think about how to make my way in life, there fell into my hands a copy of Sir Arthur Eddington's Gifford lectures [1.1]. A single sentence, but an exciting one (in his Chapter 10), lit in me the desire to become a scientist:

> *'All authorities seem to agree that at, or nearly at, the root of everything in the physical world lies the mystic formula $pq - qp = ih$.'*

One formula to understand the universe! How exciting! That should not be beyond me! But Eddington had cheated a little, for now that I understand it, the universe is still a bit of a puzzle. But he had inspired me, and he became one of my heroes. Life without heroes is a bore, and I soon acquired others; I have indicated a hero for each chapter. Suffice it to say that I shall be very content if I can do for you, without cheating, what Eddington did for me!

In this exposition it is not all frustration and regret that we are so ignorant! There are lighter moments and historical sidelights to cheer us up. And of course there is satisfaction at what has been achieved. But we should admit that there are limits to what we can assert with confidence, even though these are not always noted. Fortunately, between the scientifically known on the one hand and the scientifically uncertain, inaccessible and doubtful on the other, lies a magical borderland. It is worth knowing for its own sake, for in it flourish practically all real human delights; and they are not easily analysable by science: generosity, romance, beauty and love.

1.3 Intuition

In contemplating the universe and the physics which governs it you may well feel that you have been dropped into the middle of a jungle without a compass—lost in surroundings which are far removed from everyday experience. This is where intuition can help.

Using intuition and no mathematics I aim to take you on a journey to the limits of at least some scientific knowledge; when we finally get close to the borders of the 'jungle' we will glimpse views of discoveries yet to come and will be able to throw light on the many gaps in our knowledge. Let this book act as a compass on this journey. The idea of using intuition is that it should enable you to actually 'feel' relationships which are absorbed into the bones, as it were, using physical insight instead of mathematics. The students, the teacher, and indeed everybody, finds this to be hard, but greatly rewarding. It moves intellectual connections closer to the plane where you *understand* things. Here science comes closer to poetry and induces a genuine sense of wonder. Even a mathematically inclined person can profit from this approach. By dropping mathematics he or she may feel that this is like 'riding without a horse'. I would assure them that it is more than that. I have one warning: intuition is not enough to create new physics (which we do not actually need to do in this book). To achieve this, intuition must be coupled to good experimental and/or mathematical know-how.

1.4 Incompleteness

Now to the red thread. There is hardly any part of the scientific enterprise which can be filed away as fully 'understood'. There is always another question which stimulates further thought, more discoveries are made or new restrictions are found. Further, the theories underlying what is known from experiment are always provisional and approximate.

We thus have a 'rule of incompleteness' which says that when presented with a theory of a part of reality, you will always find failures or incompleteness provided you look hard enough. Focus on these spots, and you may find interesting new results. This new rule of thought must eventually take its place along with already famous rules: that you should treat others as you would have them treat you; and the rule of dialectics that, when there are two opposites, it is rewarding and intellectually stimulating to look for a synthesis. The new rule adds to these and brings out the 'dynamics of science'.

Is all this really needed? It is, if we recall recent suggestions that the opposite situation holds true in science [1.2, 1.3] or even in other fields [1.4]. These ideas are stimulating. But many scientists would not agree when it is suggested that the great giants of the past, who have given us not only relativity, quantum mechanics and cosmology, but also logic, calculus and the study of chaos, have made such a good job of it, that the things which are left to discover [1.5] in science are either pretty dull or too hard. We shall find little support for these views in this book.

1.4.1 An absence of fit

So there is a graininess in our description of the surrounding world, rather as we find in a television picture or on a photographic film. If you look hard enough, you will often find that something is missing. This phenomenon reveals itself in rather diverse and sometimes surprising ways. However, it is fascinating to find it. It makes you realise that scientific theory and experiment are often incomplete or imperfect. But make no mistake: they usually work well enough.

As an abstract statement it is not surprising that there is a mismatch between the world 'in itself' and our understanding or description of it—philosophers told us long ago that they are not the same: the language we use is not always appropriate. Thus the notion of position and velocity as applied to a particle becomes fuzzy in quantum theory, when applied to one particle at one instant.

The *second* purpose of this book, the 'red thread', is of interest by virtue of the detailed examples which one encounters in seeking ultimates, but often finds incompleteness and imperfection.

1.4.2   Types of imperfection

Of course everybody who is engaged in creative work looks for imperfections with a view to improving his or her creation. However, the imperfections mentioned above are not always of this type. We may be stuck with them and they cannot be removed easily or by the stroke of a pen. At best they will be removed as science takes its course over many decades. But as science marches on, new gaps in developing knowledge appear, while some old gaps may be filled.

The imperfections seem to come in three types:

(i) Intrinsic imperfections. Science itself may give us limits to what we can know. For example, given a starting point, what is the final state of a chaotic system (Chapter 5)? What are the highest and lowest temperatures that can actually be reached (Chapter 2)? It does not look as if we shall ever know. This is intrinsic incompleteness.

(ii) Limit-imperfections of theory. A hard look at scientific concepts may show that certain restrictions are not needed, or that they are unrealistic or artificial. For example, the Periodic Table is not fixed once and for all, but can be greatly expanded (section 3.4). Some theories utilize 'isolated' systems, but closer scrutiny shows that these cannot actually exist (section 4.1). These are removable, i.e. temporary, imperfections. The law of thought mentioned in section 1.4 above follows: given a scientific result, theorem or picture, see what you can discover by looking hard at the conditions of its validity.

(iii) Imperfections due to lack of knowledge. These are important since there is always a hope that they will be removed reasonably soon. There may be a problem because of missing data which are, however, likely to be supplied in the future. For example, is there a Higgs boson (Chapter 3)? Does Newton's gravitational constant change with time (Chapter 8)? Why is there practically no antimatter in the observed universe (section 7.9)

The broader questions: what is the origin of life? what is the nature of consciousness or of the brain? are even more basic. Our difficulties here arise from the innate complexity of the phenomena themselves, and, if real understanding is to arise at all, it can be expected only after decades of investigation.

These types of imperfection will be encountered often in this book, but will not normally be distinguished from one another. Do not worry if you cannot yet understand the following more advanced, and so far unanswered, questions:

- Which cosmological model is most appropriate (section 7.3)?

- The numerical values of many physical constants cannot be explained theoretically (section 8.5).

- Infinities occur in physical theory, e.g. at the big bang, and cannot be readily handled (section 8.4).

- Our understanding of irreversibility and entropy increase is still incomplete (section 4.4).

- First causes have a place in theology, but cannot be handled by science (section 9.4).

There are two more general points worth making.

(i) Scientific results are always approximate. So in some sense they are always wrong! That is why there are clever scientists who improve our understanding and make theories more nearly right. Whatever is wrong in current science acts as a spring that encourages people to advance the subject. But we will never reach an end. 'The end of science?' is a question which, in this author's view, has 'No' as the simple answer. We pursue completeness: she is an attractive, though elusive, lady. We are engaged on a quest for elusive completeness!

(ii) To see the work of a scientist in a broader background, consider the difference between scientists and, say, artists. Artists make their individual contributions: their architecture, their paintings, their sculpture remain as witnesses of their work. Scientists, on the other hand, drop their contributions into a river of knowledge which moves on and on, though their names may occasionally survive in history books, street names and possibly in the inventions that arose from their work. So we see that the pleasure in pursuing science derives for many scientists from the work itself, from the good it may cause to be done, and only for some of them from the attributes of influence and power which may result.

1.5   Human aspects

The mathematical sophistication and complication in some of the arguments of physics can lead to exaggerated claims, which have to be withdrawn later. Some 'theorems' which were part of the physics literature for decades furnish examples which will surprise even the experts (see section 6.6.1). This is one of the reasons why intuitive understanding is so important: it acts as a check on current ideas, and on complicated mathematics, and it serves as a springboard for new advances.

Research can be a cut-throat activity pursued by intelligent and ambitious people. Some always want to get there first, achieve power and/or publicity from their research and its presentation; figure 1.1 gives a humorous illustration. To attain this aim they may present a distorted picture. This is just human nature and the general public must be made aware of it, and then make allowance for it. But for others, including this author, research can be an outcome of teaching. If you teach carefully, research follows naturally. It does not follow necessarily of course, but the prerequisites are there. Cut-throat competition is best left to those who like it. I shall have more to say on this in Chapter 10.

Figure 1.1 Paul Klee 1903: Two people meet; each judges the other to have a higher position in life. © DACS 1999

1.6 Reasons for reading this book

Why should anyone want to read this book? A good reason is to get some feeling for modern scientific arguments and ideas in a reasonably compact form. Remember:

> '*...one great use of a review, indeed, is to make men wise in ten pages, who have no appetite for a hundred pages; to condense nourishment, to work with essence, and to guard the stomach from idle burden and unmeaning bulk.*'
> *Sydney Smith (1771–1845) in a 1824 review of Jeremy Bentham's* Book of Fallacies.

Each chapter in this book covers topics which have themselves been the subject of books.

This is in addition to readers possibly profiting from my emphasis on incompleteness by interpreting it at a personal level. For it seems to me that you can apply the lessons of the ubiquity of imperfections to help in your attitude to your own life. If a much loved friend, relative, politician dies, one seeks out the remaining evidence of his or her life: The photographs, the books, the houses he or she built, the cupboard he or she made. So we create mausoleums, cemeteries, memorial lectures, societies named after well-known and well-loved individuals. The spirit of the dead is thus retained in some sense, adapted to a new time and a new purpose. It cannot be retained fully. Here, too, we have to come to terms with the elusiveness of our drive for completeness. Again, unhappiness due to thwarted ambition is another aspect of a pursuit of elusive completeness. No chairmanship of a committee? Not even membership of it? No lottery win? No civil honour? These things, while perhaps of importance in people's lives, are peripheral to our work here. So let me merely emphasize that what a study of science reveals in this book is seen to be a general trend in human thought. The realization of this point can and should be an aid or solace in our personal lives.

Physics will continue to change in the third millennium. But the topics discussed here will stay relevant and remain as a crucial ingredient of whatever the new physics will bring. To keep abreast the reader is referred to the excellent science journals *Nature*, *Physics World* and the American journal *Science*.

1.7 Arrangement of the chapters

It is helpful in discussing the arrangements of the chapters of this book to distinguish between the 'macroscopic'—objects of the size of a person or a mountain—and the 'microscopic'—objects which are so small that they cannot be seen with the naked eye. For ease of understanding, it is sensible to start with the macroscopic: ourselves and the environment (Chapter 2), and only then to describe, almost as if we were doing taxonomy in botany, the microscopic: chemical elements, atoms and quarks (Chapter 3). That is different from discussing the ultimate *theory* (so far) of microscopic physics, which is the quantum theory (Chapter 6). As it is more difficult, it is postponed to a later stage. In between are chapters which help you to understand how the microscopic components make up and affect the macroscopic world (Chapters 4 and 5). Eventually you will want to know how it all links up with the *very* large, namely the universe (Chapter 7). The concluding chapters (Chapters 8 to 10) are needed to round off our appreciation of the nature of the universe and of incompletenesses, for questions of happiness and of God cannot, with honesty, be avoided.

Chapter 2

There is no free lunch
Temperature and energy:
science for the environment

2.1 Introduction

Imagine 'temperature' as the first rung of a ladder in learning about science. As we ascend it, we shall learn more about the interest of science. It is a simple start, for we all know about temperature: we take our temperature when we think we may be ill, we check the weather forecast and likely temperature forecast before we go out for a weekend. The more ambitious readers may say 'How unexciting!'. But they would be very wrong. This book will show that as you look deeply into physical processes, unexpected and exciting vistas invariably open up.

We shall use temperature, a concept everybody knows, to gain an understanding of heat and energy and to proceed from there to the science of heat, called 'thermodynamics'. This science has several laws which are important, of one of which the writer C P Snow (later Lord Snow) said, in a famous lecture on the relation between the arts and the sciences, that every well-informed person should know it [2.1]. That law is the called the 'second law'. To know something about it should be as important as knowing a few quotations from Shakespeare.

Its importance is more than just cultural. As the physics of the 20th century grew out of that of the previous one, thermodynamics was heavily used to yield quantum theory, explained in Chapter 6. Quantum theory then explained many of the early results about atoms and molecules, which we shall deal with in Chapter 3. Coming to relativity, it was a great surprise to physicists that thermodynamics turned up yet again, this time in connection with the study of black holes (section 7.8).

2.2   How cold can we get?

Human life requires a body temperature confined to quite a narrow range, normally about 36 to 41 °C or 97 to 106 °F. Daniel Gabriel Fahrenheit (1686–1736) of Dantzig lived most of his life in Holland and made the first reliable thermometer. Another thermometric scale is named after the Swedish astronomer Anders Celsius (1701–1744), and it enables us to introduce here the idea of a 'graph', giving the relation between the two scales. In our case (figure 2.1) it is simply the straight line shown. The vertical scale gives the number of °F, while the horizontal scale gives the corresponding number of °C. You can see very simply that the range of reasonable human blood temperatures in °C (36–41 °C) corresponds to a range in °F (97–106 °F). The simple increase of increments on one temperature scale with the increments on the other scale, as represented by the straight line, is called 'proportionality'.

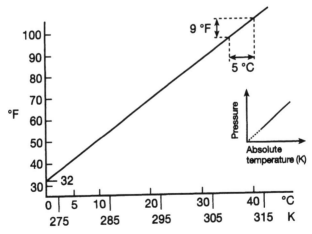

Figure 2.1 A graph relating °F to °C. The inset indicates the pressure–temperature dependence of a dilute gas.

Several gases when kept at a constant volume show another proportionality: the pressure they exert on their containers decreases linearly with temperature. It therefore drops to zero at a very special temperature. If you draw this straight line and continue it to zero pressure, you find the *absolute zero* of the temperature scale. Of course, if the gas is steam, we know that it turns into water and later into ice as the temperature is lowered. But never mind—the straight line I am talking about comes from the gaseous part and is then continued as in the inset of figuré 2.1. Fortunately you come to the same zero point, at −273.15 °C, for most of the *dilute gases*, and this explains the use of the word 'absolute'. These limiting cases are also referred to as *ideal gases*.

A third temperature scale is obtained by shifting the centigrade scale so that absolute zero actually occurs at the zero point of this new scale. This is therefore called the *absolute* or *thermodynamic* scale. The temperature of a body on the absolute scale, its 'absolute' temperature T, is denoted by T K (K stands for 'degrees Kelvin'). The unit is named after William Thomson (1824–1907) who proposed it (1848) and who joined the peerage as Lord Kelvin in 1892. I shall normally use this scale.

The size of a typical degree is the same on the Centigrade and on the absolute scale. However the Fahrenheit degree is smaller, as can be seen from the curve. There are international meetings which discuss the calibration of thermometers and temperature scales, just as there are such meetings for other measurement devices. They ensure that measurement procedures and scales are internationally agreed. There is incompleteness in thermometry below 0.65 K on the current scale called the International Temperature Scale 1990 (ITS-1990), see [2.2].

In a gas the particles (or molecules) are flying around at random, bumping into each other and into the walls of the containing vessel. At 303 K (i.e. 30 °C) their speed is about 440 metres per second, i.e. 1000 miles per hour. At lower temperatures, say at −20 °C, the speed has dropped to about 400 metres per second, or 900 miles per hour. In fact, as in the case of steam, gases tend to liquefy (water!) and later become solid (ice!) as they are cooled. An interesting aspect of this effect is that this motion does not cease completely at the lowest temperatures. This brings in the notion of *energy*.

To get an idea of energy, suppose you heat an electric kettle until the water boils. A certain amount of electricity is needed. To do the same with two kettles, you need twice the amount of electricity. To throw a ball up one needs a certain amount of effort; to throw two similar balls up, one needs twice the effort. These are examples of the energy that is needed to achieve some end. From energy let us pass to the notion of *zero-point energy*. This occurs because molecular motions tend to characteristic values at the lowest temperatures. The energy of motion, surprisingly, does not vanish at the absolute zero of temperature!

What is energy then? It is difficult to give a simple general definition. It always stands for a capability of bringing about change. If you have a gas isolated from its surroundings then, upon returning to equilibrium after stirring, its pressure and temperature may change, but its energy remains constant.

There is something elusive about energy. For example, it does not have the solid common-sense qualities of weight, speed or temperature. Weights are measured every day in the grocer's shop in grams and kilograms, and speed on car speedometers in kilometres per hour. But how do we measure energy? There is no simple 'energy meter'. There is instead the electricity meter: the bill, you remember, mentions kilowatt hours. There is the gas meter, etc. The diet experts talk about food values in terms of calories. All these quantities refer to energy. It clearly comes in a great variety of forms.

The philosophy underlying this book bids us ask: will man's attempt to reach lower and lower temperatures, in order to investigate the properties of materials at these extremes, go on for ever, or is there some limit? The answer is that it is a basic law of nature that the absolute zero of temperature, i.e. 0 K, can *not* be reached by any method. This unattainability is essentially the *third law of thermodynamics*. (For convenience of exposition I shall not consider the laws of thermodynamics in numerical order. I shall come to the other laws shortly.) It was largely pioneered by Walther Nernst (1864–1941; Chemistry Nobel Laureate in 1920). (In this book I shall use NL to denote a Nobel Laureate.)

Absolute zero can in principle be approached ever more closely. Our knowledge is incomplete because we cannot say *how* closely.

Certainly temperatures as low as one millionth of a degree K have already been reached. In the course of doing so, many completely unexpected 'low temperature' phenomena are encountered (see p 162).

2.3 Historical notes on thermodynamics

Our efforts so far have now earned us the right for the little diversion offered in this sub-section.

Box 2.1 History of thermodynamics.

Be warned that we now encounter a new incompleteness: history is never complete! In the words of Richard Feynman (1918–1988; NL) [2.3]:

'...what I have just outlined is what I call 'a physicist's history of physics', which is never correct. What I am telling you is a sort of conventionalised myth-story that the physicists tell their students...'

My story is also a myth-story, but I have made it as accurate as I can.

The development of thermodynamics took place in the age of steam engines and the search for more efficient engines was one of the motivating forces for engineers such as Sadi Carnot (1796–1832) and scientists such as Helmholtz (1821–1894), Clausius (1822–1888) and Nernst who were working on thermodynamics. Another was Joule (1818–1889) who was a student of John Dalton's (1766–1844) in Manchester, where statues of both of them now stand. Joule determined how much mechanical energy is needed to warm a given mass of water by 1 °C, and a unit of energy has been named after him.

Nernst was also the inventor of an electric lamp based on a cerium oxide rod and he interested a large German firm (AEG, Allgemeine Elektrizitats Gesellschaft) in it, although the lamp required some preheating each time it was switched on. Nernst

---

demanded, and obtained, a lump sum of a million marks instead of royalties [2.4, 2.5]. It made him a wealthy man, although his lamp lost out in the long run in competition with the Edison lamp. He told the story that when Edison (who at the end of his life held an unsurpassed number of U.S. patents (1093)) complained to Nernst about how little the AEG had paid for the patent rights for his (Edison's) lamp, Nernst shouted into the old man's ear trumpet: 'The trouble with you, Edison, is that you are just not a business man'.

In [2.5] Glasstone's *Textbook of Physical Chemistry* is cited. It is the 1946 (2nd) edition. Book reviews convey something of the flavour of science. So I recall in figure 2.2 an amusing review of its first edition by the late E A Guggenheim, Professor of Chemistry at the University of Reading, and well-known for his sharp book reviews.

Text-book of Physical Chemistry. By SAMUEL GLASSTONE. (Macmillan, 1940. Pp. 1289. 42s.)

In a certain city a councillor with grandiose ideas conceived the curious desire of having a picture of the whole city painted on a single canvas. The city indeed contained many fine modern buildings as well as older buildings of historical interest, of which good pictures already existed. But these did not satisfy the councillor, who insisted that he should have one complete canvas of the whole city; every building large or small, ancient or modern, beautiful or ugly had to be included. It is surprising that any painter could be found to attempt the task. Eventually, however, a certain travelling artist, who happened to be paying a visit to the town councillor, undertook the task of painting the whole city on a single canvas. He used a very large piece of canvas and he worked hard for two whole years, painting so many buildings each day and even working over weekends. At last his task was completed and true to his word every building, large or small, ancient or modern, beautiful or hideous, useful or useless was included on his canvas, but the product of his toil was not a picture because he had omitted to pay any regard to perspective.

The name of the city was Physical Chemistry and the name of the painter Samuel Glasstone.

E. A. G.

Figure 2.2 Book review from *Transactions of the Faraday Society* **38** 120 (1942).

2.4   What is the highest temperature?

What about the search for higher and higher temperatures? Stellar interiors can easily reach one hundred million degrees Kelvin, even though the surface temperature of our sun is 'only' about 6000 K, while its core temperature reaches several million degrees. For these high temperatures the difference between Centigrade and Kelvin, being only 273 degrees, can be ignored. A new element of incompleteness now arises since there presently exists no generally accepted upper temperature limit.

In table 2.1 I give some sample temperatures, starting with low ones and proceeding to unimaginably high ones. I have added the maximum temperature proposed in the 1970s by R Hagedorn. Its status as a maximum temperature is not as certain as that of absolute zero as the limiting *low* temperature. It arose from studies of strongly interacting gases of certain elementary particles, which gave rise to more and more particles as the temperature and the energy of the system was increased. A maximum temperature was postulated in order to limit the number of particles which can occur in the theory, and was found to be in fair agreement with experiments performed to check its value [2.6]. It still plays some part in present-day theories. In 1966 Academician A D Sakharov actually proposed an even higher maximum temperature (table 2.1), but it has not been used extensively.

Just as low temperatures furnish exciting new physics (p 162), high temperatures, too, occur in intriguing fields: the theory of the *Big Bang* and of *stellar nucleosynthesesis* and *stellar evolution*.

Let us think about extreme values for a moment. When a variable can change continuously up to some maximum (or down to some minimum) value, it is not surprising if that value turns out to be experimentally inaccessible. Why? The reason resides in the universal occurrence of small fluctuations in normal physical quantities. Thus if these values were accessible, a small fluctuation from them would take them beyond this value, and this value would then no longer be the maximum (or minimum) value. Thus one might reasonably expect that, if there are maximum or minimum values, such values cannot themselves be reached experimentally [2.7]. This gives us an intuitive understanding of the third law. What is perhaps unexpected is that there is nothing corresponding to it for the highest temperatures.

Table 2.1 Some approximate temperatures in K.

Lowest temperature reached in a laboratory	one millionth
The background radiation, a relic of the Big Bang	2.7
Liquid helium at 1 atmosphere of pressure	4.2
Coldest recorded outdoor temperature on Earth (−88 °C)	185
Freezing point of water at 1 atmosphere	273.15
Hottest recorded outdoor temperature on Earth (58 °C)	331
Boiling point of water at 1 atmosphere	373
Melting point of gold	1335
Filament of incandescent light bulb (highest home temperature)	2900
Solar surface	5800
Solar interior	ten million
Nuclear fusion reaction	hundred million
Interior of hottest stars	thousand million
Universe one second after the Big Bang	hundred thousand million
Hagedorn maximum temperature	2 million million
Sakharov maximum temperature	hundred million million million million million

2.5   What is energy conservation?

I shall now discuss the first law, not only because it is important, but also because we shall then be able to savour (in later sections) certain exceptions to it—for example by virtue of what is called the quantum mechanical uncertainty principle (section 6.2).

It is an elementary observation that if two bodies at different temperatures are put into contact, they will reach a common temperature which is intermediate between the two original temperatures. Of course you have to wait long enough! When things do not change anymore, we can say that the two bodies are in *thermal equilibrium* with each other. Why is the resulting temperature intermediate between the two initial temperatures? Why does one body not impose its temperature on the other? The reason is that heat passes between them, and what one body gains, the other loses. This heat is a form of energy, and energy is conserved for a system isolated from all outside influences. That is essentially the *first law of thermodynamics*, also referred to, in more scientific language, as the law of *conservation of energy*. Thus there can be no *perpetual motion* machine which can produce motion, and hence mechanical work, without limit from what is in fact an inadequate supply of energy. This is called a *perpetuum mobile* ('of the first kind', see section 2.7), and it cannot exist.

Great and (almost) exact as this conservation law is, it was evolved only by a painfully slow process to which the greatest scientific minds of the 17th, 18th and 19th centuries contributed. It is an impressive thought that the very term *kinetic energy*, which means the energy of motion, is only about a hundred years old, yet today every science sixth-former knows it, and, if pressed hard, might even volunteer a simple formula for it. I here just mention the main contributors, who were: Newton (1642–1727), Leibniz (1646–1716), the great Dutchman Huygens (1629–1695) and the celebrated French school of mathematicians from Descartes (1596–1650) to Comte Joseph Louis Lagrange (1736–1813). It all started with the study of energy and energy conservation in mechanics. Thereafter, in the 19th century, experimental work also embraced the sciences of heat and electricity, where new discoveries were waiting to be made. The energy conservation principle was thus broadened to include heat energy. Its further extension to include other phenomena became generally accepted in the physics of the mid-19th century. It included energy of deformation arising, for example, when you blow up a balloon, which is called elastic energy. Later chemical, electrical and magnetic energy were also included. Thus the word *energy* has a very wide interpretation and now includes, of course, atomic energy.

Inventors down the ages have tried to devise perpetual motion engines which, once set in motion, will go on indefinitely without any driving force. Magnetic devices have been proposed, turbines have been suggested with blades which are lighter than the fluid through which they move, and so on. A list of such proposals has convinced sceptics that human ingenuity knows no bounds. But to no avail! Eventually there emerged a scientific law which states categorically that these devices cannot be made. This is the law of the conservation of energy noted above; it advises the prospective inventor to seek his fortune in different fields. This is a pity, since it would be a great boon to have energy available without burning fuel or using other rare resources.

Energy conservation is used throughout science. Biologists use it in the study of photosynthesis, engineers in the design of steam engines, astronomers in discussing the origin of the heat which can be supplied by the sun. It certainly holds for these phenomena. We occasionally come across serious attempts to violate the law in some sense. For example, in the steady-state model of the universe (see section 7.3), matter is continuously generated, in violation of this law, yet this picture was a serious contender for many years. For a discussion of various unsuccessful attempts to get around energy conservation see [2.8].

2.6   A marriage of energy and mass

Let me next explain the idea of *momentum* of an object, because, like energy, it is subject to a conservation law. The pressure exerted by a billiard ball on the side of a billiard table arises from the fact that it bounces off the wall. At the bounce the ball changes its direction from going towards the wall to leaving it. This is a change of momentum. Similarly, we feel pain when a ball hits us on the neck. The faster it travels, the greater the pain. If we used a heavier ball, the pain would be greater. This pain is due to the momentum given up by the ball to our neck. In fact, momentum is proportional to both mass and speed (of the ball in our case) and its associated direction is that of the motion involved. Let me put this differently: a graph of momentum against mass is a straight line, like figure 2.1, if the velocity is fixed; similarly a graph of momentum against velocity is also a straight line, if the mass is fixed.

Let us summarize the several conservation laws of mechanics. (i) The 17th century furnished *momentum* conservation. Momentum, as we saw, is the property which a molecule of a gas carries and which is responsible for the pressure it exerts as it collides with the container. (ii) The 18th century yielded conservation of mass, and (iii) the 19th century added conservation of energy.

It is not really justifiable to attach these laws to the names of any one scientist, but the contribution of Antoine Laurent Lavoisier (1743–1794) to the study of mass and its conservation, and that of Benjamin Thompson (later Count Rumford, 1753–1814) to the study of heat and its conversion to other forms of energy, were of great significance. The latter's contribution resides in some famous cannon-boring studies which he conducted in his position as Inspector General of Artillery of the Bavarian army. Heat, which was widely believed to be a material fluid at the time, could be produced to an unlimited extent in the boring experiments. This conversion of mechanical work into heat energy led Rumford to the prophetic remark (1798)

> '...*anything which any insulated body...can continually be furnished without limitation cannot possibly be a material substance, and it appears to me...quite impossible to form any distinct ideas of anything capable of being excited and communicated in these experiments except it be* motion.'

It is indeed now considered that the heat energy of any material resides in the energy of motion of its molecules, and the above view is an early hint in that direction. He is our *hero* for this chapter.

The prominence thus given here to Lavoisier and Rumford enables us to approach the contributions made at the end of the 18th century (with only a moderate simplification of history!) by telling a story of how a marriage was arranged between mass and energy. A marriage? In what sense, you may ask. Let me explain.

A typical chemical reaction is that two molecules of hydrogen and one molecule of oxygen form two molecules of water. The hydrogen molecule consists of two hydrogen atoms and the oxygen molecule consists of two oxygen atoms. The water molecule has two atoms of hydrogen and one atom of oxygen, so that no atoms have got lost. This is just an example to remind us of chemical reactions. It is typical of many chemical reactions and in fact Lavoisier had inferred mass conservation for chemical reactions in 1789.

Box 2.2 Anne Lavoisier and Count Rumford.

1789 was the year of the French Revolution, and Lavoisier was a high government official connected with the collection of taxes. It was largely for this reason—for people at that time had no great fondness for tax collectors—that he died on the guillotine in the year 1794, in spite of his international scientific reputation. The law of conservation of mass, however, survived.

And so did his charming wife, although she was also an aristocrat. She had married him at the age of 14, had helped him with the translation of scientific papers and with the illustrations of his famous *Traité de Chimie*.

Around about this time there arrived in London a dashing young major from Massachusetts, by the name of Benjamin Thompson. America had declared her independence (1776), and Thompson, who, as a Loyalist, had helped the British, found it prudent to leave. Thompson was to become a great scientist and a practical one. He designed kitchens and lamps, studied chimneys and how to keep houses warm. Thus he used his many ingenious ideas for the scientific improvement of life; and he then used them further to gain an entry to the great establishments of England, Bavaria and, as we shall see, France. For his work in Bavaria which included the institution of soup kitchens, a recipe for Rumford soup, and the construction of the English Garden in Munich, he became a Count of the Holy Roman Empire. He is now best known as Count Rumford (see for example [2.9]). Handsome and six feet tall, he was soon a social success in Paris.

Anne Lavoisier who had, seven years before, lost on the guillotine both her husband and her father, had been irrepressible. She was, after all, one of the richest and most fashionable ladies in Paris. So Rumford was bound to meet her as he moved freely in the salons of Paris, and at the age of 48 he fell in love with her. He had married a widow before, and he was to do so again. Rumford's arrogance, however, had made him unpopular in certain quarters, and the Literary Tablet of London remarked that this 'nuptial experiment' enabled Rumford to obtain a fortune of 8000 pounds per annum—'the most effective of all Rumfordising projects for keeping a house warm'. Sadly the marriage was no great success on a personal level, but there were joined in matrimony two names associated with the conservation of energy and of mass when Anne Lavoisier became Anne Lavoisier de Rumford. The place was Paris and the year 1805. A Platonic echo was to be heard 100 years later when relativity furnished us with the equivalence of mass and energy: $E = mc^2$, where c denotes the velocity of light.

2.7 Perpetual motion?

If you want to convert heat into *work* (another form of energy) you meet another incompleteness: the conversion can be achieved only partially. For example, suppose that an expanding gas pushes a piston which then does mechanical work as in a steam locomotive. The temperature of the gas is maintained during this process by contact with a hot body, which is called a heat reservoir for this purpose. To get the gas ready for the next 'stroke' of the engine, it has to be compressed and the heat generated thereby has to be removed by contact with a cold reservoir. This not only returns the gas to its original temperature but, in addition, to its original volume. The net gain of work is of course the difference between the work done by the engine and the work of compression. You see that for this conversion of heat into work one needs *two* heat reservoirs.

Box 2.3 Sadi Carnot.

The engine described is essentially the one which was proposed in 1824 by Sadi Carnot of France (1796–1832). He was the son of the Republican War Minister and uncle of a later president of the French Republic. He died of cholera.

If the reservoirs are very large, their temperatures are practically unchanged by a heat transfer. The cycles can then be repeated again and again, and in each cycle the mechanical work produced is found to be at most a fraction of the heat energy extracted from the hot reservoir. This fraction is always less than unity and can very reasonably be called the *efficiency* of the engine. If it were unity for some engine, then let us take the oceans and deserts of the earth as hot reservoirs. During the working of the engine these would be cooled, putting an almost unlimited supply of energy at man's disposal for conversion into useful work, and without any loss. All our energy worries would evaporate! However, a generalization from our experience is that such machines are impossible. This is part of the second law of thermodynamics: the conversion efficiency is always less than unity. Even if energy conservation (first law) is satisfied, heat still cannot be converted completely into work! Wilhelm Ostwald (1853–1932; NL in Chemistry 1909) called such a machine a '*perpetuum mobile* of the second kind'. A *perpetuum mobile* of the first kind, which is also impossible, violates the first law (energy conservation), as we have seen (p 17).

The second law has another most important component. It says that any *isolated* macroscopic system has a thermodynamic variable which either stays constant or increases with time. Such a variable could be used to give an indication of the lapse of time! It is called *entropy*. We shall learn in Chapter 4 that it is one of the few variables of physics which can be universally used as such an indicator.

We have mentioned an isolated system. This is one which can exchange neither energy nor matter with its surroundings. Sometimes we talk about a less isolated system, called a *closed* system. This can exchange energy, but not matter, with its surroundings. The least constrained system is an *open* system, which can exchange both energy and matter with its surroundings. These definitions will be used later. Note also that the heat-to-work conversion efficiency of an engine is reduced by friction and by similar losses. Thus, the human mind weaves beautiful patterns which nature just fails to exhibit.

Thermodynamics governs the use of many of the engines by which we seek to influence our environment and to extract useful energy from it. This energy can be of various forms: mechanical, electrical, electrochemical (as in batteries), photobiological (as in growing plants), photovoltaic (as in solar cells), etc. Thermodynamics tells us the maximum work we can extract, and it can be used to optimize the efficiencies of any specific conversion process. Steam power plants are actually about 40% efficient [2.10] and can be based on various *thermodynamic cycles*. However, all conversion methods reject heat energy and produce pollution, car exhausts representing just one example.

Box 2.4 The laws of thermodynamics.

Here is a somewhat light-hearted summary of the three laws: the first law says that you cannot get anything for nothing; the second law says that you can get something for nothing (namely complete conversion of heat into work), but only at absolute zero; the third law implies that you cannot get to absolute zero. Now to a superficial history, attributed to Nernst [2.4] and illustrated by the portraits in figures 2.3 and 2.4. There were three main personalities whose work led to the first law: J R Mayer (1814–1878), H von Helmholtz (1821–1894) and J P Joule (1818–1889). Two people, S Carnot (1796–1832) and R Clausius (1822–1888), were the main pioneers of the second law, while only one person, W Nernst (1864–1941; NL 1920), was involved in the original statement of the third law (1907). It would appear, therefore, that nobody can formulate a fourth law!

Figure 2.3 Men of thermodynamics (Rumford, Carnot, Kelvin, Joule, Clausius).

Actually, a zeroth law is sometimes discussed and states simply that if two systems are in thermal equilibrium with a third system, they must be in thermal equilibrium with each other. It sounds obvious, but sometimes even the obvious is worth stating.

Figure 2.4. Nernst (right) and Lindemann (later Lord Cherwell) in Oxford in 1937 (from [2.4]).

2.8   Energy for mankind

A chapter on energy would be incomplete if it did not contain remarks about an important current preoccupation, namely the energy consumption on this planet. I shall therefore make a few remarks about this topic.

The earth receives energy from the sun and radiates energy into cold space, but the average temperatures of both remain roughly constant in a human lifetime. (The solar output has been found over the last 20 years to change by less than 0.1%.) Re-radiation from the earth is impeded by carbon dioxide in the atmosphere, which increases due to the combustion of fossil fuels. This has led to warnings that the temperature of the earth may increase with possibly disastrous consequences due to the melting of the ice caps. This 'greenhouse effect' is currently under investigation and it is added to by the production of other gases produced by human activities: methane, nitrous oxide and CFCs (chlorofluorocarbons).

The position is that in the period 1979–1995 air temperatures measured at the surface of the earth have risen by 0.13 K per decade. However, temperature measurements have also been made from space by the satellite Microwave Sounding Unit and have come up for the same period with a cooling trend of 0.05 K per decade. These apparently opposing results are currently being reconciled [2.11], and yield a probable warming trend. The incompleteness of our knowledge in this respect is widely acknowledged.

Another reason for care in the use of fossil fuels is that they represent a finite resource. The earth is about 4.6 thousand million years old, and only in the last three to four hundred million years have deposits of petroleum oil, shale and natural gas been accumulating. Coal deposits developed only during the last two hundred million years, since land plants and trees came late. With the industrial revolution, say during the last one hundred and fifty years, and the development of steam engines, cars, aeroplanes, etc., these valuable deposits are being used up at a tremendous rate. What took millions of years to create is being used up within a few centuries! Or, changing the time scale and regarding the earth as a middle aged person of 46 years, the fossil fuels were deposited in two years and are being used up in a few minutes!

In order to grasp mankind's energy consumption without getting involved in large numbers and strange units, let us divide the world consumption by the world population so that we shall deal with energy p.c. (per capita). Recall the primary energy recoverable from the earth: hydraulic power, crude oil, natural gas, biomass, etc. The consumption per annum is an energy consumption rate, which one has when light bulbs are burning. So let us make one continuously burning 60 watt light bulb per head of population our unit, which we shall denote by 'B' [2.12]. We then find that the primary energy demand p.c., averaged over the earth, has increased from year to year, indicating a rise in the standard of living, even though the world population also increased:

Year	1960	1970	1980	1990
Population in thousand millions†	3.02	3.70	4.45	5.29
Primary energy consumption p.c. (in B)	24.3	31.5	35.0	36.8
Electricity consumption p.c. (in B)	1.44	3.41	3.52	4.16

Electricity consumption is a secondary use of energy since it is obtained from primary sources such as hydroelectric or solar or nuclear power. Its use has clearly risen more rapidly than the consumption of primary energy.

Box 2.5 A song of two light bulbs.

If we move around normally, we give off heat and take in food, all roughly equivalent to two 60 watt light bulbs burning continuously. A Sunday school hymn gets close to it:

'Jesus bids us shine like a pure clear light
like a little candle burning in the night.
In this world of darkness Jesus bids us shine
you in your small corner, and I in mine.'

Exactly! Only substitute two light bulbs for one candle!

As this chapter is to some extent devoted to the environment, a brief mention is in order of the various primary fuels: oil, natural gas, coal, nuclear, hydroelectric, 'traditional' (use of dung, wood etc.). We also have 'new' renewable sources of energy (solar, wind, geothermal, ocean) [2.13]. However, let us consider how much electricity one

† Raw data can be obtained from [2.13]. We have used that for a population of six thousand million, one hundred million million million Joule p.a. = 8.79 B. For a brief review see [2.14].

could hope to generate if one were to cover the deserts of the earth with *solar cells*, which convert radiation directly into electricity without moving parts and therefore without lubrication etc. This *semiconductor* unit produces a current as soon as it is exposed to solar radiation. Many pocket calculators use this method. But to ask what area of desert exists on Earth is an ambiguous question. What degree of aridity do we have in mind in defining 'desert'? Do we include the cold deserts of the Arctic in the north and/or Antarctica in the south? Our language is not precise!

The experts tell us that we can take the area of 'hot' deserts as 25 million square kilometres, which is 17% of the land area and 5% of the total surface area of the earth. Then, assuming a mean insolation (averaged over day and night and the seasons) of 135 watts per square metre and a conversion efficiency of 1%, one finds that the power produced is equivalent to 94 60 watt bulbs p.a., burning continuously, for a population of six thousand million people. This would be adequate, but is of course fraught with the difficulties of distributing the electricity to the centres where it is needed. This calculation does show, however, that the energy resource residing in the solar energy intercepted by the earth is considerable. The actual conversion efficiencies of solar cells are in the range of 10–20%, depending on the materials used. We took here a mere 1% to allow for other uncertain losses. It is desirable to improve solar cell conversion efficiencies, and crucial to lower their manufacturing costs. A brilliant new idea here would be very important for mankind (and might attract a Nobel prize).

There is also incompleteness here. It refers to our current inability to make better use of the solar energy incident on the earth. It is so often wasted: the tops and walls of houses could utilize the radiation falling on them.

We should distinguish the 'high-tech' solar cells from devices depending on solar water heating, which are 'low-tech', as they do not require the sophistication of the semiconductor industry. Such solar panels for swimming pools are already economic in the UK and they are in use in favourable climates.

2.9. Summary

We are inspired in this book by the search for the limits to what we can know. Surprisingly, even as far as a simple concept like temperature is concerned, we have found such limits. They can only be approached, but not reached. Examples are: (i) a lowest temperature, at which, furthermore, the molecular velocities in a gas do *not* cease (see p 12), and (ii) an ideal, but unattainably high, conversion efficiency from heat to useful work (see p 22). This is important since the supply of energy is always limited (section 2.7). We have also noted the law of energy conservation (p 18), which has, however, tiny exceptions.

Thus we have come across what the mathematician E T Whittaker called 'principles of impotence' (see p 218). We cannot reach absolute zero, but we do not know how close to it we can get. We cannot convert heat into work without loss, but do not know how close to it we can get. These are spurs to further progress. We are similarly aware from the newspapers of the greenhouse effect and the danger to the ozone layer, but the precise extent of the expected damage is not known.

A box of gas can have the size of a human being. Such objects are called *macroscopic*. Thus temperature and pressure are typical macroscopic variables. In the next chapter we shall consider the nature of atoms and ask about their constituents. We are then in the field of *microscopic* physics.

Chapter 3

Painting by numbers
Elements and particles:
science as prediction

3.1 Introduction

This chapter is devoted to the arrangements of components in a table. First we have the chemical elements arranged in the Periodic Table; thereafter we have the elementary particles arranged in simple figures of six or eight sides. My idea is to show how the work of today is inspired by the thoughts of yesterday, and how intrinsic incompleteness is encountered in the case of the Periodic Table as well as in the case of elementary particles.

Box 3.1 Mendeleev and Meyer.

The year is 1887. The British Association for the Advancement of Science meets at Manchester amidst cigar smoke and conversation. One interest in the meeting of Section B (Chemistry) is that two internationally famous chemists are present: Mendeleev (1834–1907) from St Petersburg in Russia and Lothar Meyer (1830–1895), first of Karlsruhe and later professor in Tübingen in Germany. It was not so easy to travel long distances in those days, and people were anxious to hear, as well as see, the two famous chemists. After the President's address, there was a call for a speech by Mendeleev. His English was not good. So he declined and merely rose from his seat to bow to the audience. Almost at once Meyer rose from the seat next to Mendeleev. As if to avoid any mistakes, he remarked 'I am not Mendeleev.' Quiet pause. 'I am Lothar Meyer.' There was a round of applause to show that the disappointment at being unable to hear Mendeleev was at least partially replaced by hearing Meyer. He asked in perfect English to be allowed to speak in German. He then expressed on behalf of Mendeleev and the other foreign chemists the pleasure they had derived from the Presidential address.

Who were Mendeleev and Meyer and what was the reason for their fame?

3.2   Chemistry in 1867

We must go back to 1867. A bright young chemist, Dimitri Mendeleev, 33 years old, had just been elected to the chair of chemistry at the University of St Petersburg. He prepared his lectures with such enthusiasm that they were a great success. The auditorium was filled, men from other departments came to hear him, as if some great drama was enacted each time he spoke. He had a great deal to speak about for much chemistry was known at the time. *Elements* were known as substances which cannot be decomposed chemically into simpler substances, and elements like oxygen, carbon, gold and silver were, of course, well known from the times of the alchemists; and so were mercury and sulfur and their combinations. In fact about 63 elements had been discovered. These elements could combine to form thousands of *compounds*—for example, water consists of hydrogen and oxygen. A very fundamental transformation must take place when water is formed in this way, since its properties are so different from those of its constituents. The same applies to most compounds.

For many of the elements and compounds boiling points, melting points, density, colour and many other items had been studied. The science of chemistry has also a rich ancestry in alchemy, an early form of chemistry associated with magic and the conversion of base metals into gold. Man had asked himself for a long time how gold could be made, for it had been regarded as the perfect metal since 4000 or

5000 BC. Man had asked himself, too, which chemicals to take to relieve pain, ensure long life and, of course, beauty. Thus alchemy was a forerunner not only of chemistry but also of medicine. People were, of course, also keen to make gold in a kind of *transmutation*. It is ironic that, in spite of all the errors of alchemy, it was quite correct to look for ways of changing one element into another artificially. The dream of the alchemists was realized when in 1919 Rutherford converted nitrogen to oxygen.

So you can imagine how a keen young professor of chemistry, Mendeleev, was practically overwhelmed by the information he had to impart: there was no general system; the facts did not hang together. Some materials could lie around for a thousand years without change, (platinum, gold); others would even eat through the wall of the vessel in which they were stored (fluorine). Some were light (hydrogen), some were abundant (oxygen), some were useful (iron). Even the colour of an element or compound could change on heating, and so could its density. There were just no guiding scientific principles.

Scientists believe in order, and if it is not obvious, then we seek to uncover it. Restricting oneself in the first place to elements, there was one number known that proved to be very useful to Mendeleev. That was the number of times the atom of an element was heavier than an atom of hydrogen. This is now called the *atomic weight* (see box 3.2). Atomic weights were determined by John Dalton (1766–1844) and others from a careful study of the weights of the participating elements in chemical reactions. Reacting elements like hydrogen and oxygen do not give a molecule of water with no material left over, as if one were mixing two paints to get a new colour. The fit is much more precise, as in a jig-saw puzzle. Thus it is exactly two molecules of hydrogen which react with one molecule of oxygen to give two molecules of water. From the measured weights of reacting elements the atomic weights were obtained. Thus oxygen was found to have atomic weight close to 16. Lead, a very heavy atom, is about 207.2 times heavier than an atom of hydrogen, etc.

The atomic weight has a great advantage over other numbers, since it remains characteristic of the element whether it is in liquid, gas, or solid form, whether it is ground to dust or in the form of a large crystal. All atoms of an element were furthermore believed to be exactly like each other. So you could take an extended chessboard of 63 squares, each representing an element, and write on each the known characteristics, including the atomic weight, and try to arrange them in some sort of order. Could you sensibly place them on the extended chessboard so as to reveal similarities and differences? Let me now steal a forward look at the structure of atoms, as this will make the discussion easier.

Box 3.2 The Bohr atom.

In a rough model a typical atom is now known to consist of a positively charged *nucleus*, with much lighter negatively charged *electrons* circulating around it so as to make it electrically neutral. The nucleus itself consists of positively charged *protons* together with electrically neutral *neutrons*. Each of these two particles is about 1833 times as heavy as an electron. This model is rough, being an old but picturesque model associated with Niels Bohr (1885–1962; NL 1922), the famous Danish physicist. One might expect the atomic weight of an element to be close to a whole number, namely the number of protons plus the number of neutrons in the nucleus, the mass of the electrons being so small that they can be neglected in a first approximation. But it is usually not a whole number because of the contribution arising from the interaction energy between all the particles involved. Recall from the equivalence of energy and mass (Box 2.2) that such energy must be reflected in the resulting mass.

The number of protons in an atom, however, *is* a whole number and is called its *atomic number*. The chemical properties of an atom are determined largely by its electrons and their number. But this still leaves the possibility of different numbers of neutrons. As the number of neutrons is changed, you arrive at different so-called *isotopes* of an atom of a given chemical species. These different isotopes are still given the same chemical symbol, e.g. H for hydrogen. *Outside* an atom, neutrons will live for an average of 10 minutes before decaying into an electron and an anti-neutrino (p 59).

3.3 The Periodic Table and three predictions

To illustrate Mendeleev's procedure, take 13 of his 63 elements, and let us try to arrange them. Of course we shall make it easy for ourselves by taking elements which are now known to belong together as regards their physical and chemical characteristics. Here they are then, together with their rough atomic weights (not their atomic numbers!), as known in Mendeleev's time:

boron (B) 11, carbon (C) 12, nitrogen (N) 14, *,
magnesium (Mg) 24, aluminium (Al) 27.4, silicon (Si) 28, phosphorus (P) 31,*,
zinc (Zn) 65.2, arsenic (As) 75, *,
cadmium (Cd) 112, indium (In) 116, tin (Sn) 118, antimony (Sb) 122.

First note the gaps in the atomic weights, as indicated by asterisks. This suggests that you might be able to rewrite the elements in a better table:

					Part of row
	B 11,	C 12,	N 14	2
Mg 24,	Al 27.4,	Si 28,	P 31	3
Zn 65.2,	*	*	As 75	4
Cd 112,	In 116,	Sn 118,	Sb 122	5
Part of 'group' II	III	IV	V	

The elements in each 'group' have closely related properties, though in recent tables group II may be split up (p 36). We have used here some chemical knowledge about these elements. Each element now shares a column with other elements which have roughly similar reactivity with oxygen (which was a well-studied kind of reaction). By writing row and group numbers in the margins of the table above, a modern terminology has been introduced (see table 3.1).

There are clearly two gaps, denoted by asterisks, which we can call the aluminium-plus-one element and the silicon-plus-one element. The Sanskrit word for 'one' is 'eka', so Mendeleev called the apparently missing elements eka-Al and eka-Si, and predicted their existence. He was correct in this prediction, and these elements are now called gallium and germanium.

Mendeleev went further and predicted several properties of the new elements, such as the chemical formulae for their oxides, chlorides and fluorides, their densities and their boiling points. This was indeed 'science as prediction' par excellence. Mendeleev's table had a third new element: eka-boron (atomic weight 45.1, also in row 4, see section 3.4). With the aid of the table he also suggested errors in some of the experimentally determined assignments of atomic weights.

The table was refined during the next few years, and in 1882 Mendeleev and Meyer were jointly awarded the Davy medal of the Royal Society for this work. The medal was only instituted in 1877 for 'recent discoveries in Chemistry made in Europe or North America'. Meyer's contributions were considerable, but he was not as explicit and detailed in his prediction of new elements. This is where Mendeleev scored. At a time when the discovery of each new element caused a stir among scientists, he foresaw three: not by experiment, but by detailed study of the evidence already there. Thus the guidance his table provided enabled you not only to classify existing elements but also to look for new ones.

3.4 Confirmation

What happened in the years between the revised Periodic Table of 1871 and the Manchester British Association meeting of 1887 with which we started? The papers by Meyer and Mendeleev lay on the shelves and were ignored, until something happened in 1875.

In order to describe it, let me introduce a possibly well-known term: *frequency*. It is the number of times a phenomenon repeats itself exactly in a second. Thus a low note in music corresponds to a low frequency. A high note corresponds to a high frequency. This is easy to remember! An analogous distinction applies to light. The colour of a light wave corresponds to a certain frequency. For example, red yields a lower frequency than blue. Now each element has a kind of *fingerprint* which resides in the different frequencies of the light it can be made to emit. An unfamiliar fingerprint leads to the conclusion that a new element has been discovered.

Now, in far away Paris, a new element was discovered in the Academician Wurtz's laboratory. There was excitement in France as the sealed envelope containing details of the discovery was opened in the

French Academy—the usual procedure to prevent theft of new ideas. As Gallia is the Latin name for France, the new element was called gallium and was reported to be much like aluminium. In due course the news reached Mendeleev: there was no radio, and news travelled slowly. He at once wrote to the French Academy that the element was eka-Al of atomic weight approximately 68 and *specific gravity* about 5.9. (The specific gravity is simply a measure of how many times the material considered is heavier than water.)

The discoverer, Lecoq de Boisbaudran (1838–1912), had found a lower specific gravity, but Mendeleev advised him to try again with a purer specimen, and in due course there it was: more or less as predicted! The scandinavians came next in 1879 with the discovery of a new element they called scandium, clearly eka-B (L F Nilson, 1840–1899). The Germans (C A Winkler, 1838–1904) then found a new element they called germanium in 1886. It was found to have almost the exact properties of eka-Si, as predicted by Mendeleev, who thus becomes the *hero* of this chapter (figure 3.1). Thus Mendeleev's three 'ekas' had been found in three different countries, each eager to get its name onto the periodic table. Today silicon, gallium and germanium are important elements for the manufacture of silicon chips and for the semiconductor industry generally.

Thus law and order were established among the chemical elements. No need to look for an element lying between In 116 and Sn 118 for example, for the Mendeleev table did not allow it. Easy to start your lectures now with the lightest element, hydrogen, and take the elements in ascending order of atomic number, or you could study them group by group (table 3.1). On the right-hand side we find the *permanent gases* (helium, neon, argon, etc), to their left we have the metals (e.g. copper, silver, gold). The permanent gases could not easily be liquefied at the time—hence their collective name.

It was of course far easier to discover naturally occurring elements. Thus helium was first discovered in the sun (helios is the Greek word for 'sun') by its characteristic spectrum. In fact, the *noble* (or *permanent* or *rare* or *inert*) gases were all discovered in the last century. This is shown below, where the atomic number is also given:

2 helium (1895); 10 neon (1898); 18 argon (1894);
36 krypton (1898); 54 xenon (1898); 86 radon (1900).

Table 3.1 The Periodic Table around about 1996. The numbers in front of the symbols of the elements denote the atomic numbers. The numbers at the top are the atomic weights.

1	2	3	4	5	6	7	8	9	10	11	12	13	14	15	16	17	18
1.00797 1H Hydrogen																	4.0026 2He Helium
6.939 3Li Lithium	9.012 4Be Beryllium											10.811 5B Boron	12.01 6C Carbon	14.01 7N Nitrogen	15.999 8O Oxygen	18.998 9F Fluorine	20.182 10Ne Neon
22.99 11Na Sodium	24.31 12Mg Magnesium											26.98 13Al Aluminium	28.09 14Si Silicon	30.97 15P Phosphorus	32.06 16S Sulfur	35.45 17Cl Chlorine	39.95 18Ar Argon
39.10 19K Potassium	40.08 20Ca Calcium	44.96 21Sc Scandium	47.90 22Ti Titanium	50.94 23V Vanadium	52.00 24Cr Chromium	54.94 25Mn Manganese	55.85 26Fe Iron	58.93 27Co Cobalt	58.71 28Ni Nickel	63.54 29Cu Copper	65.37 30Zn Zinc	69.72 31Ga Gallium	72.59 32Ge Germanium	74.92 33As Arsenic	78.96 34Se Selenium	79.91 35Br Bromine	83.80 36Kr Krypton
85.47 37Rb Rubidium	87.62 38Sr Strontium	88.91 39Y Yttrium	91.22 40Zr Zirconium	92.91 41Nb Niobium	95.94 42Mo Molybdenum	(98) 43Tc Technetium	101.07 44Ru Ruthenium	102.9 45Rh Rhodium	106.4 46Pd Palladium	107.9 47Ag Silver	112.4 48Cd Cadmium	114.82 49In Indium	118.69 50Sn Tin	121.75 51Sb Antimony	127.6 52Te Tellurium	126.90 53I Iodine	131.30 54Xe Xenon
132.91 55Cs Cesium	137.34 56Ba Barium	Lanthanides 57-71	178.49 72Hf Hafnium	180.95 73Ta Tantalum	183.85 74W Tungsten	186.2 75Re Rhenium	190.2 76Os Osmium	192.2 77Ir Iridium	195.09 78Pt Platinum	196.97 79Au Gold	200.59 80Hg Mercury	204.37 81Tl Thallium	207.19 82Pb Lead	208.98 83Bi Bismuth	(210) 84Po Polonium	(210) 85At Astatine	(222) 86Rn Radon
(223) 87Fr Francium	(226) 88Ra Radium	Actinides 89-103															

Lanthanides →

138.91 57La Lanthanum	140.12 58Ce Cerium	140.91 59Pr Praseodymium	144.24 60Nd Neodymium	(145) 61Pm Promethium	150.4 62Sm Samarium	151.96 63Eu Europium	157.25 64Gd Gadolinium	158.9 65Tb Terbium	162.5 66Dy Dysprosium	164.9 67Ho Holmium	167.3 68Er Erbium	168.9 69Tm Thulium	173.0 70Yb Ytterbium	175.0 71Lu Lutetium

Actinides →

(227) 89Ac Actinium	232.04 90Th Thorium	(231) 91Pa Protactinium	238.03 92U Uranium	(237) 93Np Neptunium	(242) 94Pu Plutonium	(243) 95Am Americium	(247) 96Cm Curium	(247) 97Bk Berkelium	(251) 98Cf Californium	(254) 99Es Einsteinium	(253) 100Fm Fermium	(256) 101Md Mendelevium	(253) 102No Nobelium	(257) 103Lr Lawrencium

For our purpose the table itself is of greater significance than the details of the elements and their abbreviations. Just as fingerprints are used to identify people, the spectral fingerprints of atoms can be used to identify them uniquely. Note that for many years the Periodic Table ended with uranium, which has atomic number 92.

Figure 3.1 Picture of Mendeleev from the Faraday Lectures of the Chemical Society 1869–1928. (Reproduced courtesy of the Library and Information Centre, Royal Society of Chemistry, Burlington House, London W1V 0BN, UK.)

Box 3.3 The periodic law today.

Mendeleev was now a very famous man, even though he was never elected to the Academy in Russia. The periodic system became so firmly entrenched in the minds of scientists that today it is taken for granted. So much so that often modern popular science writers [3.1–3.6] do not mention it. Whitehead [3.7] devotes exactly one sentence to it. Yet the 'periodic law', as Mendeleev called it, has great philosophical significance, being the forerunner of today's tables of elementary particles and of the great international competition to predict new particles and to find them experimentally. For recent historical remarks see [3.8].

3.5   The atom in the 1890s

In 1873, at the British Association meeting in Bradford, the Scottish physicist James Clerk Maxwell (1831–1879) observed: 'The mind of man has perplexed itself with many hard questions. Is space infinite, and if so in what sense? Is the material world finite in extent, and are places within that extent equally full of matter? Do atoms exist, or is matter infinitely divisible? The discussion of questions of this kind has been going on ever since men began to reason, and to each of us, as soon as we obtain the use of our faculties, the same old questions arise as fresh as ever' [3.9]. These questions are indeed still part of our search for completeness. At the British Association meeting in 1894 Lord Salisbury, in his Presidential Address, remarked similarly: 'What the atom of each element is, whether it is a movement, or a thing, or a vortex, or a point having inertia, whether there is a limit to its divisibility, and, if so, how that limit is imposed, whether the long list of elements is final, or whether any of them have any common origin, all these questions remain surrounded by a darkness as profound as ever' (cited in [3.10]).

Box 3.4 The energeticists.

At a famous confrontation in Lübeck, Germany, the notion that atoms exist was advocated strongly by the well-known Austrian physicist Ludwig Boltzmann (1844–1906), but it was opposed by the so-called energeticists, who did not believe in the atomic structure of matter; instead they preferred to regard nature as merely macroscopic, since atoms could not be seen or felt. They included Wilhelm Ostwald and Georg Helm. Helm's letters to his wife bear eloquent witness of the intensity of the debate [3.11]. On 17 September 1895 he wrote from Lübeck (my free translation)

'Dear Elise! The great event has now taken place. The talk was, I think, quite successful. There was applause and praise, but in the discussion things were pretty tough. Boltzmann commenced with friendly and appreciative remarks... but then he started to attack both mine and Ostwald's papers. He, and later Klein, Nernst, Oettingen, touched on matters... for which I was not prepared... Ostwald and Boltzmann attacked each other quite strongly... the hall was half-full so that several hundred people witnessed the discussion...'.

He explains that he visited a restaurant after the session in order to calm down.

The matter was of course resolved in favour of atoms and against the energeticists. In fact, with the discovery of *Brownian motion* and its interpretation as due to the buffeting of pollen particles by the molecules of the liquid in which they are suspended, the reality of atoms was eventually accepted. This was a success for Boltzmann's ideas. It is an ironic quirk of history that he had committed suicide just before this success became widely known. The reason for his suicide has remained a matter of conjecture.

The cut and thrust of debate in science continues to this day, and modern versions of Helm's letters are probably still being written today. Ostwald, and also Ernst Mach (1838–1916), a fellow Austrian and originally strongly opposed to Boltzmann's atomic ideas, were later both reconciled to the atomic hypothesis. Mach wrote Boltzmann's obituary in a Viennese daily paper. He made a relevant remark there, namely that while a sensitive nature is required to do creative scientific work for oneself, robust nerves are needed to take in other people's contributions. He attributed some of Boltzmann's depression to this circumstance.

There is a conglomeration of ideas going back to about 1880 and known as (Ernst) Mach's Principle. They imply that when we have to push in order to move matter, i.e. to overcome its inertia, we are actually doing work against effects due to the matter in the surrounding universe. Thus an inertial frame (see p 155) is determined by the averaged motion of distant astronomical objects, as Bondi puts it in his 1952 book (see [9.10] below). It is a good reference frame, since bodies not acted upon by forces are at rest in it or move with constant speed in a straight line [3.12].

3.6   The atom split

The Mendeleev predictions were to become the prototype of other predictions. The basic philosophical question, which had once occupied the Ancient Greeks, was this: if atoms, i.e. indivisible units of matter, had mass, how were you to think of the bits of matter which made up these atoms? Some people thought that this was a meaningless question, as these smaller bits could not be isolated. Also people had no idea if atoms were structureless or not. When, therefore, J J Thomson (1856–1940; NL 1906) announced at a Friday evening discourse at the Royal Institution in London on 30 April 1897, that he had found a negatively charged particle whose mass was less than one thousandth of the mass of an atom of hydrogen, people wondered where it was to be placed on the Periodic Table, thinking this was a very small and light atom. Such was the success of the Periodic Table in the eyes of scientists that they found it difficult to think of a new particle in any way other than that it must be a new kind of atom. Was it perhaps the primordial atom of which all other atoms were made?

So J J Thomson produced a paper for the 1899 British Association meeting in Dover entitled 'On the existence of masses smaller than the atoms'.

The new particle could be stripped off different atoms. It was always the same: light, fast and with the same mass and negative charge. It was not a new atom or the primordial atom, it was *part* of an atom, and is now called an *electron*. Atoms had been split!

Thirty years later, in the 1930s, three elementary particles were known in terms of which the elements could be explained (p 32):

- the relatively heavy, positively charged particle near the centre of the hydrogen atom, called the proton;

- the electron (negative electric charge; about 1833 electron masses were found to equal the mass of a proton);

- the neutron (approximately of proton mass, but electrically neutral);

- a fourth elementary particle was also known: the neutral constituent of radiation called the photon. It is believed to be massless.

There was more fun to be had with electrons, as we shall see (p 64).

3.7 Incompleteness

It is interesting now to look for the imperfections. The most important one is that the Periodic Table, while sound in principle and not upset by the discovery of the electron, is incomplete:

(a) A serious limitation arises from the fact that more elements can be created artificially, adding to those already present. We have now gone far beyond element 92 (table 3.1). The atomic number and the year of discovery of the newer elements is given below:

93 neptunium (1940); 94 plutonium (1940); 95 americium (1944); 96 curium (1944); 97 berkelium (1949); 98 californium (1949); 99 einsteinium (1954); 100 fermium (1954); 101 mendelevium (1955); 102 nobelium (1957); 103 lawrencium (1961).

Recently, elements 104 to 110 have been made in the heavy-ion accelerator in Darmstadt, Hesse, Germany. In 1996 element 112 with 165 neutrons was discovered there, but it breaks up rapidly after manufacture. In the Dubna Laboratory of Heavy Ion Reactions, Russia, they found element 114, and now elements 116 and 118 have been found at the Lawrence Berkeley National Laboratory in the US; these are the 'superheavies' and the search proceeds!

Here are some proposed names for the more recently discovered elements [3.13]. They were agreed only after intense international competition:

- 104 rutherfordium (after Lord Rutherford)

- 105 dubnium (after Dubna where the relevant Russian experiments are performed)

- 107 bohrium (after Niels Bohr)

- 108 hassium (after the state of Hesse, Germany)

- 109 meitnerium (after Lise Meitner, 1878–1968).

Lise Meitner made important contributions to our understanding of nuclear fission, and her name is remembered, for example, in connection with the Hahn–Meitner Institut in Berlin–Dahlem.

We learn a simple lesson: the Periodic Table is never going to be completed, though the 'stable' elements form an important subset.

(b) Many of the elements can exist as chemically similar species, which are however distinct by virtue of their different atomic weights. These are the isotopes (p 32) of a given element. This is not too serious, however, as you can note down the isotopes of any element in its appropriate square of the Periodic Table.

(c) Any historical account of scientific developments has to be incomplete. There are many little steps taken by many different people so that 'history', as normally expounded, is always a great simplification. See box 2.1.

3.8 Plum-pudding or planetary system?

It took about thirty years of experiment and theory for a model of the atom to be evolved which made sense and which at the same time gave an explanation of the Periodic Table. The atoms were soon to be regarded as small planetary systems with the heavy protons and neutrons in the centre and the light electrons swirling around them. This model was not the first and most obvious one. It was largely due to Ernest Rutherford and was forced on scientists by scattering experiments which showed that the main mass of an atom was at its centre. These normal atoms are stable (i.e. they do not decay spontaneously into smaller pieces) and this was quite a problem for the classical physicists. The negative electrons would be expected on the basis of the classical theories of the nineteenth century to spiral rapidly into the central nucleus. The reason is this: an electric charge travelling in any closed orbit suffers accelerations. These are needed to keep it in its closed orbit. This is known to anyone who has been thrown to the side of a car which is cornering sharply. Things are very different for straight line motion, which can occur with or without acceleration. Anyway, the accelerations lead, in the case of charged particles, to the emission of radiation and hence loss of energy from them. Another reason for the inward spiralling is that the nucleus is positively charged, while the electron has a negative charge, and opposite electric charges attract!

The atomic model thus constructed is as open as the solar system. If we picture the cross section of a typical atom as given by a circle whose radius is 100 m in diameter, then the central nucleus could be represented by a small stone about 1 cm in size. The electrons would be still smaller as they swirl about the nucleus. But the stability of the structure was a puzzle. An alternative was the plum-pudding atom of J J Thomson, who assumed a uniform spread of positive electricity

Box 3.5 Rutherford.

Ernest Rutherford (1871–1937; NL in Chemistry 1908, knighted in 1914) and F Soddy (1877–1956) used particles (or 'rays' as they were then called) produced by radioactive substances as projectiles to obtain a transmutation as early as 1902. Cockcroft (1897–1967; NL 1951) and Walton (1903–19; NL 1951) obtained a transmutation of Li to He in 1932, now using artificially accelerated particles. Actually Rutherford (figure 3.2) gave his results the blander name 'transformation', remarking to Soddy '...don't call it transmutation. They'll have our heads off as alchemists' [3.14]. Indeed, the alchemists' dream of making gold from lesser materials was finally within reach.

Rutherford also produced his now famous theory of the scattering of particles in 1911. This is one of the reasons why it has been claimed that he was the only physicist who did his greatest work after receiving the Nobel prize [3.15]. Special attention is paid in this reference to the history of elementary particle physics from 1895 to 1983. There may, however, be exceptions since John Bardeen (1908–1991) received Physics Nobel prizes in 1956 and 1972!

Figure 3.2 Walton, Rutherford and Cockcroft in 1932.

with the electrons as the raisins. This model was at least stable, but it did not agree with the scattering experiments: plum-pudding or planetary system?

To the rescue of the planetary model came, in 1913, a very early version of quantum theory in the hands of the tall, pipe-smoking and somewhat ponderous but brilliant Niels Bohr (see box 3.2). The achievement of the planetary atom, supported by quantum mechanics, turned out to be that it was able to explain that the atoms of the 92 elements were made of structures which were 'stable', whereas they were expected to collapse pretty quickly if you used only classical physics. Further, these structures were explicable in terms of only the three particles we have already met: protons, neutrons and electrons. (Remember, though, that 'stability' here and earlier is always taken with reference to disturbances considered to be 'reasonable': a powerful beam of radiation could destroy any of them!). The planet-like electronic orbits are now called *Bohr orbits*.

The fingerprints of atoms, which I introduced on p 34, could be interpreted in the following manner. The frequency of the light emitted was regarded as due to electrons in some of the atoms dropping from one rung in an energy ladder to a lower rung. There are many transition possibilities presented by an atomic energy ladder. The totality of them, for a given atom, gives rise to its fingerprint.

Conversely, you can think of incident radiation being absorbed by an electron in one level, which is thus promoted to a higher energy level. The absorption properties of an atom can be understood in terms of such upward transitions. Thus you arrive at a pleasing symmetry: downward transitions are associated with emission of radiation and upward transitions with absorption. The theory of radiation was to utilize this symmetry in due course.

Now let us move along a row in the Periodic Table. One proton (and possibly neutrons) and one electron is added as you go to the next higher position in the table (say from K to Ca), leaving a normal atom electrically neutral. Each electron in an atom occupies what is called a specific quantum state. The rules are that an additional electron cannot then be accommodated in the same state.

That there exist different levels of energy, instead of a continuously varying energy, is known as *quantization*. This idea is now part of our normal language and is of course a characteristic feature of quantum theory. The general idea was first introduced by Ludwig Boltzmann in a different context (a 1877 study of the basis of statistical mechanics [3.16]) and more specifically by Max Planck (see p 150) in 1900. Quantization always involves what is now known as *Planck's constant*, usually denoted by *h*.

The energy carried by a wave was found to be given by Planck's constant multiplied by a frequency (and possibly also by some integer). When radiation was involved the carrier of this energy was called a *photon*. It can be regarded as a wave or as a particle, and so was called a *wavicle* by Eddington [1.1, p 199] in the early days of quantum theory. It has zero mass and charge. Similarly the differences in energy between two atomic energy levels could also be regarded as Planck's constant multiplied by a frequency. These *quanta*, or lumps of energy, do not occur in classical mechanics, where energy can usually vary smoothly. So how can you eliminate the energy gaps in the theory to recover the classical results? The gaps would disappear if you could, on paper at least, pass from quantum equations to the corresponding pre-quantum (or 'classical') equations by 'letting *h* tend to zero'. As this leads to sensible results at least in some cases, this procedure can be regarded as a new principle. It is called *Planck's correspondence principle*, since it establishes a correspondence between classical and quantum mechanics.

We know another such principle which is somewhat different. This is *Bohr's correspondence principle*, which states that classical results can be obtained when the system energy becomes large enough (or, equivalently, in the limit of large quantum numbers [3.17]). Both principles tell us that classical physics, e.g. Newton's mechanics, should be contained within the more general theory of quantum mechanics. That is of course reasonable: as science advances new ideas are expected to give a broader perspective, and to include earlier models as special cases.

To obtain more insight into the properties of electrons, let us learn next about a new quantity, namely *angular momentum*. A cyclist moving in a straight line has no angular momentum, but the wheels have, because they are turning. The earth has, because it is moving around the sun. Angular momentum arises from rotation (figure 3.3). Elementary particles can also have angular momentum, but since they have no well-defined extension, their angular momentum (called spin), though part of their nature, like their mass, cannot be easily visualized, and it has therefore been called *intrinsic*. The spin of an electron is a typical quantum phenomenon, and so is expected to involve Planck's constant; classically it is zero: intrinsic spin was not known in classical physics! The value of the spin is always a multiple of ½h, but it can be positive or negative, normally referred to as 'spin up' or 'spin down'. This would correspond, according to classical ideas, to a top spinning in one direction or its opposite. If spin is included in the specification of the *quantum state* of an electron in a hydrogen atom, then each such state can be occupied by only one electron. This is the exclusion principle of 1925, due to Wolfgang Pauli (1900–1958; NL 1945).

One caution: if we have two neighbouring hydrogen atoms, an electron in one atom *can* of course be in the *same* quantum state as an electron in the atom next door. We may then regard the two atoms as forming *one* system.

In order to complete the specification of quantum states we actually need additional integers, called *quantum numbers*: the importance to the Pythagoreans of the integers has been brought back by quantum theory! But now they are no longer associated with magic.

As atomic orbits can take only a limited number of electrons, new orbits are created as electrons and protons are added to an atom. The heavier atoms have a more complicated system of electronic orbits. The orbits are classified into shells and when an atom has no electron beyond a completed shell, then it is inert: the inert (or noble or permanent) gases He, Ne, Ar (see p 35) are examples: they do not react much chemically. When there is one electron outside the last completed shell, on the other hand, there is good chemical reactivity, which is governed by these electrons. Examples are the alkali metals lithium (Li), sodium (Na) and potassium (K).

When discussing elementary particles in general, a greater variety of spins opens up, and we just note it here in passing. We find that spins have a magnitude of one of two types:

$$(\tfrac{1}{2}, \quad \tfrac{3}{2}, \quad \tfrac{5}{2}, \quad \tfrac{7}{2}, \ldots)h \text{ for so-called fermions}$$
$$\text{and } (0, \quad 1, \quad 2, \quad 3, \ldots)h \text{ for so-called bosons.}$$

This shows how the notion of electron spin has been generalized. Bosons are named after the Calcutta physicist Satyendra Nath Bose (1894–1974), who died shortly after an international celebration in Calcutta of his eightieth birthday. After that celebration I shook hands with him. With his shock of white hair, he urged the attendees to keep afresh 'that wonderful spark' which gives fulfilment to scientific work [3.18]. The Indians are immensely proud of him. Fermions take their name from the Italian physicist Enrico Fermi (1901–1954; NL 1938), famous not only as a physicist, but also for his work on the atomic bomb. The two different types of particles seem to be as different as chalk and cheese, and obey different laws.

Popularly speaking, fermions are introverts, and their statistics is such as to tend to keep them away from each other. This statistical effect acts over and above the attraction or repulsion due to the electric charge the particles may have. It will be recalled that you can have positive or negative electricity, and that like charges repel, while opposite charges attract. Bosons, in contrast, tend to come together; they are extroverts.

In the fairly recent *string* theory, however, which is not yet generally accepted, an imagined space of eight dimensions is utilized, instead of just the three dimensions of ordinary space plus the one dimension of time. Operations in *this* generalized space enable you to transform fermions into bosons and conversely. At this level of abstraction fermions and bosons *could* thus be considered to be of the same type. With some additional assumptions (which may be loosely referred to as *supersymmetry* or *SUSY* for short) you would then expect a supersymmetric partner for every boson and fermion (see p 54). But this is going further than we need to in this book, since these partners have not so far been found experimentally.

When you look at the Periodic Table with our new insights, it tells you rather directly that: the periodicities among the elements can be regarded as the results of Bohr orbits according to the 'old' quantum theory of Bohr and others, and later understood even more clearly in terms of the 'new' quantum mechanics of the late 1920s (sections 6.2–6.4).

Before we leave the Periodic Table, let us take aboard three thoughts: (i) the exclusion principle, spin and existence of well-separated energy levels result from quantum theory. Classical physics did not prepare us at all for these surprises. (ii) Is it not remarkable that, wherever you go in the universe, and at whatever time, you will not find an infinite variety of substances, but only those few, ninety or so, normal elements and their compounds? Many of the simpler chemical reactions can be understood in terms of the electronic structure of the participating atoms. (iii) In some real sense quantum theory has reduced chemistry to physics.

Figure 3.3 Albert Einstein with angular momentum. From Tullio Pericolo, *Woody, Freud and Others* (Munich: Prestel, 1989). I am grateful to Dr John Sweetman, University of Southampton for drawing my attention to this artist.

3.9   A taxonomy of particles

Our endeavour to understand the nature of matter has been going on for millennia and thousands of people are involved in these studies at the present time. Here I give a simple introduction.

At the bottom there lies a philosophical puzzle already recognized by the Greeks: how can you understand matter in terms of indivisible units (called atoms from the Greek word 'atomos', meaning indivisible), when the units themselves have extent and mass? Does the constituent mass not itself have units of mass from which it is constructed? Our story does indeed reveal the validity of this basic incompleteness in our understanding by passing from atoms to protons to *quarks* (discussed below). Are further steps to be expected? *Strings*, so small that it would take one hundred million million million of them to pass from one end of a proton to the other, are new units which represent a current hope.

Let us divide the history into five steps, each revealing more underlying structure. They are not meant to be strictly chronological.

I. Atoms. From the Greek ideas of atoms to the Periodic Table. This takes us from Democritus in about 585 BC to 1896 AD. We have dealt with this in the preceding part of this chapter.

II. Atomic structure. The picture of an atom as consisting of a nucleus and electrons started with J J Thomson's discovery of the electron (1897) and ended up with atoms as a solar system in miniature (Bohr and others). This period ends in about 1927 because the 'new' quantum mechanics was then developing. That electric charges come (excepting quarks) as multiples of the electron charge, and that the proton charge is of exactly opposite sign to the electron charge are currently not fully understood. Of the three types of incompleteness noted in Chapter 1, this represents a presumably temporary imperfection.

III. Antimatter. In 1928 Dirac (1902–1984; NL 1932) predicted the existence of what has turned out to be an antielectron. This is a particle with the same mass and spin as the electron, and it is now called a positron. In 1932 it was discovered in cosmic rays, which enter the atmosphere in large numbers from outer space, by C Anderson (1905–19; NL 1936). Every particle is now known to have an antiparticle with the same mass and spin but of opposite electrical charge (and opposite colour charge; see p 57). Some electrically neutral structures, like the photon, are their own antiparticles. It follows that the radiation from antimatter is effectively indistinguishable from the radiation emitted by ordinary matter. If antimatter existed in the universe in the proximity of normal matter, but not in excess of it, it would annihilate with matter and produce a great abundance of photons. This may have taken place in the early universe so that there would be little antimatter left. One argument in favour of this idea is that cosmic rays contain only traces of antimatter. The fact that antimatter does not appear to occur naturally in any quantity in the universe, while it can be made by man e.g. in the Large Electron–Positron Collider (LEP), still awaits an agreed explanation. In fact, Dirac in his 1933 Nobel lecture suggested that the universe is symmetric in particles and antiparticles. That would be elegant, but this view is not currently accepted. There are occasional unconfirmed reports of the sighting of larger systems of antimatter. It is a topic which is not fully understood.

Actually, a device to detect antimatter (the Alpha Magnetic Spectrometer (AMS)) has been designed and will soon be sent on a space flight, possibly in the year 2001. It may determine if antimatter presently occurs naturally in the universe. If we do not find antimatter, we will conclude that it was produced only in the Big Bang, but quickly annihilated in explosions due to its interaction with ordinary matter. This would not, of course, be a final proof that antimatter does not now exist naturally and in reasonable quantities, e.g. in the form of antistars. There may be, for example, an anti-you somewhere—but this notion is more appropriate to science fiction than for us in this book. We must be satisfied with the limited success arising from imperfect knowledge.

IV. Transmutation of the elements. This can be brought about by using some of the fast projectiles which are available in the physicist's toolbox and shooting them at appropriate atoms.

V. The standard model. This emerged during the period since about 1950 and it represents current ideas about subnuclear physics, as summarized in table 3.2. I give here a rough taxonomy of particles. For more background see, for example, [3.19–21].

The table is divided vertically to distinguish fermions from bosons. It also distinguishes the carriers of forces (on the right-hand side) which may exist only temporarily on their own, from the particles (on the left-hand side) which have broadly speaking a continuous existence and may be considered to be the truly elementary particles. They are the *leptons* and *quarks*. 'Leptons', based on a Greek word, refer to the lighter elementary particles. As to quarks, Murray Gell-Mann (b 1929, NL 1969) named them, following a remark by James Joyce in *Finnegan's Wake*:

> *'Three quarks for Muster Mark!*
> *Sure he hasn't got much of a bark*
> *And sure any he has it's all beside the mark.'*

Indeed, as we see from the table, three quarks is exactly what is needed, not only by Muster Mark, but also for the construction of protons and neutrons, and also for the rarer *baryons*. The composite particles (baryons and *mesons*) appear further down the table. Baryons (the name has a Greek root) refer to heavier particles. Baryons and mesons can still be regarded as 'fundamental', because they are so important; but they are not 'elementary', because they are composed of other particles, namely quarks.

There are six types (sometimes called 'flavours') of quarks. Of these the up quark is the lightest. The top quark proved to be the most elusive of all, but twelve 'sightings' were claimed in April 1994 by a team at the Collider Detector at Fermilab (CDF), in Chicago, where they have the Tevatron accelerator facility. In recent years the number of sightings has been increasing. The mass, huge by elementary particle standards, was believed to be just below that of a gold atom. Alternative names for top and bottom are truth and beauty.

The electric charge of a quark is a fraction of an electronic charge: $\frac{2}{3}$ for the up, charm and top quarks, and $-\frac{1}{3}$ for the down, strange and bottom quarks. This suggestion by Gell-Mann and George Zweig was accepted by the physics community only 15 years or so after its proposal; it had been taken for granted until then that the electron provides the smallest unit of charge.

The (uud)-structure (where u stands for up and d for down) of the proton, given in the table, gives it the electric charge $(\frac{2}{3} + \frac{2}{3} - \frac{1}{3})e = e$, the electronic charge, as it should. Similarly the (udd)-structure of the neutron leaves it with a charge of $(\frac{2}{3} - \frac{1}{3} - \frac{1}{3})e$. This is zero, i.e. the required electrical neutrality. Quarks have an additional 'charge' which comes in *three* distinct forms, so that the more familiar notion of positive and negative charge is not applicable. What can we possibly think of, to convey the idea of three types of charge, such that certain important combinations leave us without any charge? Physicists adopted 'colour' as the distinguishing marker: red, blue and some intermediate colour like green or yellow—any three will do. Quarks feel the colour force, but have not been seen individually. Particles which have been seen individually, notably baryons and mesons (see table 3.2), are always colourless, i.e. they are almost unaffected by the colour force. This feature corresponds to zero charge when we talk about ordinary electricity. Thus for baryons the three quarks have three different colours, producing a 'colourless' or 'colour-neutral' result. Similarly, mesons, which are formed of a quark and its antiquark, are also colour neutral. The need for a colour force will be made plausible under point (c), p 56.

The relationship between colours is analogous to what is seen in 'Newton's disc experiments', in which an appropriately coloured disc is rotated rapidly, and then looks white, i.e. colourless. Similarly, white light is split into various colours on being passed through a glass prism.

Returning to table 3.2, note that the bosons are on the right-hand side of the table. They include the graviton and the photon, which will be discussed more fully later. In fact, bosons mediate the four known forces: the strong or colour force mediated by *gluons*, the electromagnetic force mediated by photons, the weak force and gravitation. Gluons may in fact be defined as particles that transmit the strong force. The name is based on the notion of 'glue' as used to stick quarks together, and then it is given a Greek-sounding ending. (These names seem to require some apology!)

All fermions, and some bosons, carry their own characteristic mass, and all are subject to gravitation, as indicated at the bottom of the table, where the sensitivities of the various particles to the forces are indicated by arrows. The particles which are composed of quarks and held together by gluons are surrounded by a thick black band to indicate that they are subject to the strong force.

Table 3.2 The 'standard model': fundamental particles and their interactions.

Fundamental particles with a continuing existence; all are *fermions*.	Particles which give rise to forces when exchanged between fundamental particles; all are bosons.
They can be created or destroyed only if an appropriate antiparticle is simultaneously created or destroyed.	They may exist temporarily on their own.
	Not constant in number: may be created (or destroyed) in interactions.
For example, the neutron n decays to proton p, electron e⁻ and antineutrino $\bar{\nu}_e$ as shown on the right. This conserves the number of baryons (one, before and after) and creates a lepton-antilepton pair: $e^- + \bar{\nu}_e$.	Whereas photons are massless and carry energy for the electromagnetic force, the weak force carriers $W^\pm$ and Z^0 have enormous masses and consequently make an extremely brief (virtual) appearance.
	Thus the beta decay of the neutron, $n \rightarrow p + e^- + \bar{\nu}_e$, may be considered to be the transformation of a d quark in the neutron (n ≡ udd) into a u quark by emission of a W^-, leaving a proton (p ≡ uud); the W^- decays immediately into electron and anti-neutrino:
Particles are counted positive, anti-particles negative, so the numbers of both leptons and baryons remain constant.	$$d \rightarrow u + W^- \rightarrow u + e^- + \bar{\nu}_e$$ This shows the role of quarks and the weak force carrier W in beta decay.

Elementary fermions (all have antiparticles)		Forces and their carriers			
		Strong or colour force	Electro-magnetic force	Weak force	Gravitation
6 Leptons 3 families	**6 Quarks** in 3 colours r, g, b 3 families	**8 Gluons** carry colour- + anti-colour-charge to and fro between quarks to form observable colour-neutral composite particles	1 carrier: Photon γ mass: zero	3 carriers: $W^\pm, Z^0$ Intermediate vector bosons very heavy: carry a short-range force	? carrier: Graviton no evidence yet

e⁻	electron	u	up
ν_e	electron-neutrino	d	down
μ⁻	muon	c	charm
ν_μ	mu-neutrino	s	strange
τ⁻	tau	t	top/truth
ν_τ	tau-neutrino	b	bottom/beauty

Composite particles
(quarks bound together by gluons)

Baryons	Mesons
Baryon = 3 quarks bound by gluons: e.g. nucleons p, n, p = uud proton n = udd neutron, excited states: Δ^{++} = uuu delta Λ^0 = uds lambda Ω^- = sss omega	Meson = bound quark + antiquark Examples: pion $\pi^+ = u\bar{d}$ $\pi^0 = u\bar{u} - d\bar{d}$ $\eta^0 = u\bar{u} - d\bar{d}$ kaon $K^+ = u\bar{s}$ phi $\phi = s\bar{s}$ J/psi $\psi = c\bar{c}$

Hadrons
← particles sensitive to the →
strong interaction

Unified as the electroweak force in the 'Standard Model' by means of the Higgs boson which gives masses to the elementary fermions (quark and leptons) and to the W^+ and Z^0.

The mass of the Higgs boson is large but could be low enough for future experimental measurement.

← Involved in the electroweak interaction →

← Sensitive to the gravitational interaction →

3.10 Basic forces

The force between particles may be pictured by an interchange between them of *messenger particles*. If these have no or little mass, the force has an effect over long distances. If they are relatively heavy, the force has a short range. It is a little bit like playing a generalized form of tennis: using very heavy balls the players have to keep close together to keep the ball in play, whereas with very light balls the players can separate widely. The four elementary forces will be discussed below under (a), (b), (c), and (d). Their unification into a single generalized force is an important, but unsolved, theoretical problem. Einstein studied the unification of (a) and (b) (gravity and electromagnetism) without success. Is there a fifth force (see p 176)?

(a) Gravity. I begin with this most familiar force, which, interestingly, is normally too weak to play an important part in the study of atoms. It is assumed that the so-called *graviton* is the carrier of this force. It has not yet been discovered; this is presumably 'only' a lack-of-knowledge incompleteness. But we know, of course, that gravity binds the solar system and, through the Newtonian *theory of gravitation* and *Newtonian mechanics* can be used to predict eclipses, and is responsible for tides. It is the most universal force.

Now gravitation acts on all forms of energy, including radiation, which we already know to consist of photons. Conversely, gravitational effects are produced by any form of energy. Thus photons are acted upon by gravitational fields, and these gravitational effects increase with the energy of the photons. At very large energies the gravitational force between photons, which is normally negligible, can, remarkably enough, become comparable to the other forces.

But normally the gravitational forces between typical gravitationally interacting particles is small. The gravitational attraction which we note when we drop a cup, and it breaks, is of course *not* small. Why? Because it acts between bodies consisting of many, many typical gravitationally interacting particles! The earth is 'large' for the purposes of this consideration.

(b) Electromagnetism. The photon is the main constituent of radiation, i.e. of light, and has no intrinsic mass. Of course it has energy and comes in all colours, and also in colours, or frequencies, beyond both the red and the blue ends of the rainbow (see section 2.4). It is a messenger particle in the sense that it mediates the electromagnetic force, for example the force between electric charges. This is the binding force for atoms, which keeps the nucleus and its orbiting electrons in a stable state when they are undisturbed by external actions. It also governs chemical reactions and is responsible for electric light, radio and television.

Photons of different frequencies exist, making up the colours of the rainbow. As we have learnt to expect from what was said in section 3.8 about the atomic energy levels, the energy of a photon is proportional to its frequency and Planck's constant enters here again. Both electric and magnetic forces play a part in the generation and transfer of radiation such as radio waves and light, so that we speak of *electromagnetic* forces, etc.

Box 3.6 QED.

Here, in brief, is the story of electromagnetism, which is a unification of electricity and magnetism first brought about by James Clerk Maxwell. It leads to radio communications, e-mail and the World Wide Web. They are unthinkable without it. His equations are now part of syllabi in physics, mathematics and electrical engineering throughout the world. The theory has been expanded since Maxwell's time to take account of quantum theory. There is also a connection with relativity. The newer theory is called QED or quantum electrodynamics.

(c) The strong force. A third ('strong') force is clearly needed to hold together nucleons (protons and neutrons), as noted at the end of section 3.9. The constituents of protons and neutrons, or, more generally, of all hadrons, are the quarks, as already discussed above.

What of the glue that keeps the quarks in the protons and neutrons together, and also keeps the protons and neutrons as components of the nucleus? The colour force (see table 3.2), like the other forces we have met, has to be mediated by messenger particles and these are called gluons. They take the place of photons in quantum electrodynamics (QED), and were discovered experimentally in 1979. Because of the occurrence of 'colour', the new theory is called QCD, which is short for quantum chromodynamics. The strong force is responsible for the stability of atomic nuclei and hence of plants, animals and ourselves. It is also responsible for the possibility of producing electricity in nuclear reactors. Following a theme of this book concerning the imperfection of scientific knowledge, and the search for its completion, we must expect the J J Thomson story of 'a mass smaller than an atom' to be repeated in rumours and guesses yet to come, concerning 'particles smaller than a quark'. Why leptons and quarks both come in the form of three 'families' (see table 3.2) has yet to be fully explained.

As a matter of historical interest, an early meson to be discovered was the pion. The short range of nuclear forces led Hideki Yukawa (1907–1981; NL 1949) to predict in 1935 the existence of a particle about two hundred times as massive as the electron. In 1947 a matching (unstable) particle (the pion or pi-meson) was in fact found in cosmic rays which enter the atmosphere from outer space as high energy particles. Muons are much more numerous in cosmic rays at sea level. They were discovered in 1936.

(d) The *weak force*. Its messenger particles are relatively heavy and this force is too weak to bind particles. Its main characteristic is that it enables quarks to change into one another. This is an important property, its first manifestation was in the darkening of photographic plates stored near uranium salts due to the uranium decay products. This discovery was due to a member of the scientifically inclined family: the Becquerels. The discoverer was Antoine-Henri (1852–1908; NL 1903). One disintegration per second is a (rather small) unit of radioactive decay,and is called the Becquerel. Later, nucleosynthesis in stars, i.e. the production of heavier elements by the fusion of nuclei, and also the production of radiation by the sun, were found to involve the weak interaction. Just as electromagnetic theory combined the effect of magnetism and electricity into one theory, so much progress has also been made towards combining the electromagnetic and the weak force into one force, called the *electroweak* force. The weak force is responsible for the decay of some atoms. The term 'electroweak' was invented by Abdus Salam (1926–1996; NL 1979), who was originally from Pakistan, and who also did a great deal for theoretical physics in the third world.

The unification leading to the electroweak force was rewarded by a Nobel prize in 1979 to Steven Weinberg (b 1933), Abdus Salam and Sheldon Glashow (b 1932). The further unification with the strong force is an important research topic. The result is the *grand unified theory* (GUT). Further unification with the force of gravity into one single force leads to *theories of everything* (TOE). This objective, on which Einstein spent the later years of his life, still escapes scientists. It is now believed that that the forces remained unified down to a tiny fraction of a second after the Big Bang (one hundred million million million million million millionth of a second). The particle energies were then on average much larger than is achievable in today's particle accelerators (the factor is one hundred million million million times and the actual energy is ten million million million million electronvolts) [3.22]. A suggestion has recently been made by Keith Dienes and his colleagues at the European Laboratory for Particle Physics, CERN, near Geneva, that the forces may perhaps remain unified for longer. The term 'TOE' is actually objectionable as it can give the impression that scientists are headline-grabbing, which is bad for scientific public relations.

How many types of particles are there? Or, put more carefully, how many degrees of freedom do they have? Incidentally, everybody uses the term 'particles' nowadays, their wave properties being taken as generally understood. Here is the calculation (see for example [3.23]).

Quarks: three flavours, each with three colours and two spins:
$3 \times 3 \times 2$ 18
Antiquarks   18
Gluons: eight types, each with two spins (called helicities).   16

This makes a total of 52 degrees of freedom. If we have to go beyond the standard model, e.g. to take account of the neutrino rest mass (section 3.11), additional 'superparticles' may come into play, adding to this number. The reader need not worry—you can never get absolutely up to date, as was already known to *Tristram Shandy*, who tried to write his biography, but could never catch up.

3.11 Predictions of particles

Just as the Periodic Table enabled us to predict new elements, so the proper arrangements of known particles in elementary particle physics also led to successful predictions. We shall note six.

(i) The *neutrino*. The electroweak interaction between particles (as indicated in table 3.2) is responsible for the decay of a radioactive nucleus with the production of an electron. The electron was called the beta particle in the early days and the decay was called the beta-decay. In these decays there appeared to be a violation of the general principle that energy and angular momentum have to be conserved in particle collisions and decays. Extensive discussions of this matter led Wolfgang Pauli (1900–1958; NL 1945) to suggest in 1930 that an electrically neutral and very light particle was also emitted in the decay and that its spin was responsible for the apparent non-conservation of angular momentum. Following the discovery of the neutron in 1932, Fermi christened it the 'neutrino'—'the little neutral one'. Such a particle would be very hard to detect and Pauli wagered a bottle of champagne that it would not be found. He remarked that he had replaced something we cannot understand with something we cannot observe! He had to pay up a quarter of a century later when Frederick Reines (1918–1998; NL 1995) and Martin Perl (b 1927, NL 1995)) found it—actually they found an antineutrino. We now know of three types of neutrino and their antiparticles (see table 3.2), the tau neutrino being discovered at Fermi Lab in July 2000. Neutrinos can oscillate between the various types (for example, see [3.24]). For a long time they were believed to have zero rest mass.

A tremendous effort was made to check this out in the form of the hundred million dollar detector, called Super-Kamiokande, operated since 1996 by about 120 physicists from 23 institutions and headed by the University of Tokyo. Using a 50 000 ton water tank and 13 000 photomultiplier tubes they found evidence for a non-zero neutrino mass [3.25]. This means that the standard model has to be patched up in some way. As we know that there are a vast number of neutrinos throughout space, this finding also has important implications for estimates of the mass in the observable universe. As to the photon, everybody believes its mass really is zero. Experimentally it

has been shown that it must be less than the very tiny one millionth-
...millionth (eight times) of a gram [3.26].

Box 3.7 Wolfgang Pauli.

Pauli was an incisive and highly critical person. It was believed for a long time that, speaking rather roughly, the laws of physics are the same for a system and for the mirror image of that system. T D Lee (b 1926; NL 1957) and C N Yang (b 1922; NL 1957), however, proposed the violation of this result on theoretical grounds in 1956. A story I have heard is that Pauli bet his reputation that this law would *not* be violated. It was a fair guess, after all his faith in conservation theorems had already led him to success in connection with the neutrino. When the Lee and Yang prediction was, however, verified by C S Wu, people came to Pauli and said: 'What do you say to this?' 'Well', said Pauli, 'it just shows how clever I am. I did not put any money on it, of which I have little. I just bet my reputation, of which I have plenty!' On another occasion, when discussing a rather poor scientific paper, he is supposed to have remarked: 'It is so bad, it is not even wrong'.

In my own dealings with him, he was, however, straightforward. For 28 years or so two basic theorems of statistical mechanics (whose precise nature does not concern us here), due to John von Neumann were accepted. The argument was improved by Pauli and his distinguished colleague Markus Fierz. When my student Ian Farquhar and I found that the theorem was actually useless, we sent our paper to Pauli for comment and approval. We expected some criticism. But not at all. After a few months, and without the least difficulty, the matter was agreed and completed by letter.

The reader familiar with Goethe's 'Dr Faustus' may enjoy some verses from a play written and performed by several students of Bohr's in 1932 [3.27]. This extract is intended to whet the reader's appetite for more.

Faustus (N Bohr)... *If ever to a theory I should say:*
'You are so beautiful!' and 'Stay! Oh, stay!'
Then you may chain me up and say good-
bye-
Then I'll be glad to crawl away and die.

Neutrino (replacing Gretchen) sings to Faust:

My mass is zero, my charge is the same.
You are my hero, Neutrino's my name

(ii) The eta meson (see table 3.2 and figure 3.4). For the classification of baryons Murray Gell-Mann introduced a new quantum number s, called 'strangeness', inspired by Francis Bacon's 'There is no excellent beauty that has not some strangeness in the proportion'. If one plots s on the vertical axis of a graph and the electric charge on another, but oblique, axis, one finds a hexagon whose vertices represent allowed states of mesons. The obliqueness of the second axis is just chosen for convenience. The discovery in 1961 of the eta-meson (as a second meson, in addition to the pi-meson, for the centre of the diagram) completed the octet of mesons, and confirmed its existence. The theory was called the eight-fold way by Gell-Mann (following the idea that there are eight ways of avoiding pain according to the Buddhists). The second eta particle is heavier and was discovered later.

(iii) The *omega minus* baryon. Gell-Mann proceeded in a similar way with baryons, using a graph of strangeness versus charge, and more sophisticated methods, to show that a ten-particle system could be constructed. From this he predicted the omega minus particle in 1962. It was to occupy the position of most negative strangeness and charge, which was thus far unoccupied by any particle (figure 3.5). When the omega minus was found in 1964 the eight-fold way was thereby firmly established. The quark content of the heavier baryons (see table 3.2) is also shown in figure 3.5. Note that three up-quarks appear together on the right. But by the Pauli exclusion principle the fermions in a system must be in distinct states, and so must be specified by different quantum numbers. Hence we are led naturally to the need for another quantum number, colour, which has already been noted (sections 3.9 and 3.10(c)). By introducing it as an additional quantum number, to be specified for each baryon, the Pauli principle is satisfied. All the particles in figure 3.5 are colour-neutral, i.e. have no colour charge.

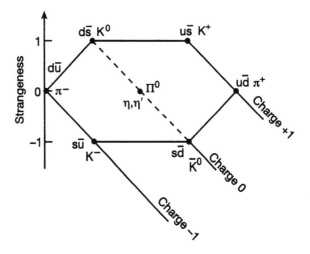

Figure 3.4 The mesons (eight particles).

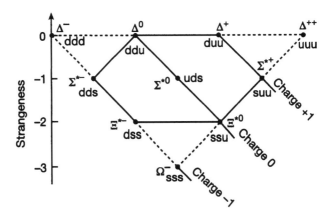

Figure 3.5 The heavier baryons (ten particles).

(iv) The *charmed quark*. Ten years later, in the 'November revolution' of 1974, the standard model (table 3.2) also became firmly established through the discovery of hadrons made of charmed quarks. Quarks have not been observed directly, but their properties can be inferred from high energy collisions. The charmed quark turned out to have a mass greater even than a proton mass, which is itself believed to be made of (other) quarks (uud, see table 3.2). This is remarkable and gives impetus to the search for the next deeper layer of matter.

(v) The *intermediate vector bosons* (W^+, W^-, Z^0) are shown in table 3.2, and were predicted in papers by Glashow (1961), Weinberg (1967) and Salam (1968). They were discovered in 1983 by a team at the Centre Européenne pour la Recherche Nucleaire (CERN) led by Carlo Rubia.

(vi) The *Higgs boson*. This particle has been predicted, but it has not yet been discovered. The Higgs mechanism is named after the Edinburgh physicist Peter Higgs (b 1929). So far we have kept secret the fact that the standard model concerns zero mass particles, and only photons and gluons are believed to be of zero mass! For the other particles the mass is put *into* the theory, instead of coming *out*, as a number, *from* the theory. This is still a big gap in the existing theory. The interaction of the fields already introduced with the stipulated Higgs field is supposed to provide a mechanism of giving masses to the particles (e.g. the W and the Z) which are known to have a mass. It is a hard experimental problem because the Higgs boson, considered as the quantum for the Higgs field, has itself an unknown mass, believed to lie between 130 and 1000 proton masses. Recent calculations favour the lower values. Around about the year 2005 it may perhaps be created and seen in a large machine (the Large Hadron Collider) of the European particle-accelerator Laboratory at CERN.

In the standard model of elementary particles, the Higgs boson remains the only particle which has been predicted but not discovered. What *is* an elementary particle, anyway? Perhaps it must satisfy the requirement that is its field must appear in the fundamental field equations [3.28]. Possibly elementary particles just represent the low energy vibrations of *strings*. These tiny one-dimensional objects are, in a recent theory, believed to replace the model of point particles. Their vibrations correspond to those of a violin string. These are obviously open questions.

3.12   Electrons yield modern electronics

The discovery of the electron had huge practical implications, which will now be traced. Electrons are known to form an *electron gas* in metals and in some other solids. Also they often interact so strongly that they lose their independence. Because particles in a liquid are closer together than in a gas, one talks then of an electron liquid. The energy of such an electron system resides to such an extent in their interaction energy, which is therefore spread throughout space, that it is no longer possible to localize it at a point in space. Such electrons, and such strongly interacting particles generally, are called *quasi-particles*, and, for the reason given, their energy is regarded as a *non-local* quantity. The mass, called under these circumstances the *effective mass*, of such electrons is changed from that of the free electrons by their interactions with each other and their surroundings in the solid. Thus they can be lighter or heavier than in a free state, by a factor which may be as high as ten or even a thousand.

For many years it was an unsolved problem to show theoretically that electronic charges must come always as a multiple of the electron charge. Then came the realization from theory that the charge on a quark can be ⅔ or (−⅓) (see section 3.9). This sort of value turned up again very unexpectedly in the 1980s in low temperature solid state research from a study of the *Hall effect*. This venerable effect, found in 1879 by E H Hall (before the discovery of the electron), consists of a transverse voltage developed across a *semiconductor* carrying an electric current through a perpendicular magnetic field.

A semiconductor is a solid in which the relevant electrons occupy two bands of allowed energy separated by an energy gap. The upper energy band is normally fairly empty so that electrons in it can gain energy from an applied electric field, producing good electrical conductivity. The lower band is normally almost filled with electrons and therefore does not contribute much to conduction: the electrons in it cannot gain energy very easily, unless of course they jump across the main gap. Thus we have the first band (the conduction band) and a second one, called the valence band, which conducts via missing electrons or '*holes*' whose effective charge is positive.

We will do a tiny calculation next. The band structure of a solid is a consequence of quantum theory and so involves Planck's constant, whose magnitude is well known. Here we take it to be of the order of one electronvolt (eV) multiplied by one femtosecond (fs). A femtosecond is tiny: one thousandth millionth millionth of a second! The electronvolt is a typical energy for electron transitions in a semiconductor. Since the uncertainty principle (section 6.2) requires that the measured time in a process must exceed Planck's constant multiplied by the energy involved, we are left with one femtosecond (fs) as typically the shortest time interval for switching or for other transitions in semiconductor electronics. This can be seen as follows:

$$\text{time interval} > h/(1\ eV) \backsim eV\ fs/eV \backsim fs.$$

This turns out to be an important constraint in semiconductor electronics. Strictly, one has to use h divided by 2π, and with modern values one finds 0.66 fs for the right-hand side; but this is a numerical detail.

Some semiconductors are '*p-type*' because they carry their current predominantly by holes ('p' stands for positive!). Now Hall effect measurements distinguished between *n-type* and *p-type* semiconductors. As everybody knows, chips and computers depend on semiconductors and are rapidly becoming more powerful. The number of transistors which can be placed on a semiconductor chip doubles roughly every year and it is also suggested that computers double in power every 18 months. These rough rules of thumb are often called *Moore's law*.

The Hall resistance, obtained from the Hall voltage divided by the current, was found in 1979 to have very surprising properties at low temperatures. It changed in steps rather than continuously, suggesting a novel form of quantization. Furthermore, it depended basically only on Planck's constant and the value of the electron charge, so it became a way of checking the value of some fundamental constants. This discovery led to the 1985 Nobel prize for Klaus von Klitzing. It confirmed, incidentally, what everybody knew by that time, that electric charge was indeed quantized. At very high magnetic fields, however, additional steps were found. This suggested that fractional electron charges, typically of one third or two thirds of the electron charge, played a crucial role. We should not think in this case of an actual particle with this charge. Rather it appears to be an artifact arising from the properties of strongly interacting components of a quantum fluid. This discovery attracted the 1998 Physics Nobel prize (Robert Laughlin). A charge of $e/5$ was found in 1999.

3.13 Summary

An important characteristic of science was explained in this chapter. By arranging chemical elements in a properly designed table, you can check that the properties of these elements change appropriately as you move from one element to the next in the table. New elements can be predicted and characteristic properties of the elements have been checked by this means. Because new elements can be made artificially, such a table can never be complete. Indeed, the search for completeness, which is a theme underlying this book, is well illustrated by this phenomenon. In the last sections of this chapter it is shown how these ideas are carried over into the physics of elementary particles.

Here the existence of certain particles has been predicted and verified from a study of particle tables, just as chemical elements have been predicted from a study of the Periodic Table. One particle at least (the Higgs boson), is still only a conjecture. Indeed the incompleteness of our understanding increasingly leads to the question as to whether it is possible to decompose the 'quarks', which are currently believed to be the basic massive particles, replacing the atoms of Greek science. This is probably an intrinsic problem of incompleteness which may never be solved. It reminds us of how J J Thomson shocked scientists at the turn of the century by the discovery 'of a particle smaller than an atom'. In a way this discovery gave rise to the electronics industry. There are other lack-of-knowledge incompletenesses of course, but they can in principle be solved. For example: does the proton decay, albeit very slowly? Has the photon a mass? Further, physicists have tried to make particles one-dimensional by treating them as tiny 'strings' and even going beyond this concept. These theories are already a going concern, but they need more work.

Between the macroscopic and the microscopic regions lie areas of knowledge which involve both: time and entropy (Chapter 4) and chaos and life (Chapter 5). Not only are these topics clearly important in themselves, but they also take further the theme of incompleteness in science.

Chapter 4

Why you cannot unscramble an egg
Time and entropy:
science and the unity of knowledge

4.1   What is entropy?

The flow of heat and its conversion into work (section 2.4) on the one hand, and the unruly wild bouncing about of atoms or molecules of a gas in an enclosure on the other hand, are different descriptions of the same system. The first uses the large-scale ('macroscopic') properties, the latter smaller-scale (atomic or 'microscopic') concepts. As one cannot normally follow the atoms or molecules individually in a large system, there being too many of them, a statistical approach is needed. The mechanics of energy transfer between individual atoms then gives way to what is called *statistical mechanics*.

We know that statistics is useful, for we use it all the time to describe, for example, changes in the public appreciation of different political parties. In that case we do not say how any one individual would vote, but come up with statistics instead. Similarly, instead of talking about individual atoms we may talk about the number of atoms in a system which have a certain property of interest. With these numbers we can do statistical mechanics (see, for example, [4.1, 4.2]).

If there are N atoms, one takes the three coordinates and the three components of momentum of each atom as the variables. They form a good set for defining the state of a system. If their possible values are divided into the smallest recognisable ranges, we have a very large (but finite) number of possible *states* in which the system may find itself. A 'state' signifies here a situation in which each component of each position and momentum variable of each particle has been assigned a (possibly approximate) value. If the specification is by quantum numbers, we refer to it also as a *quantum state* (see section 3.8).

In a true state of equilibrium each state of any given system has a definite probability. Take a gas for example. The chance that all molecules are in the left hand part of the container is small, whereas a uniform distribution of molecules throughout the container is much more probable. It is useful to invent a quantity which applies to the system as a whole, and is small in the first case and large in the second case. This is called *entropy*. A homely example follows.

Unleash a child in a playroom in which all the toys are neatly packed away (low entropy, figure 4.1),and soon the toys will be scattered all over the floor (high entropy, figure 4.2). Here we have two equilibrium states: the tidy room, the untidy room, and the transition

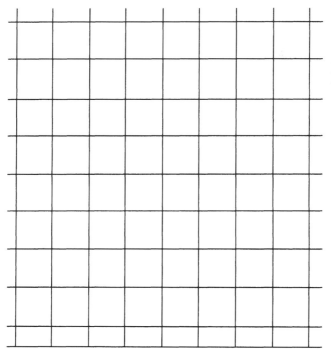

Figure 4.1 Low entropy! A piece of graph paper.

induced by an 'external influence', to wit a child. Thus in any transition to an equilibrium state the entropy of a system goes up. But just a minute! Suppose this is an intelligent child who creates *greater* order. This possibility of decreasing entropy by the exercise of intelligence brings us to the border with biological sciences (see section 5.4).

Figure 4.2 Large entropy! Kenneth Martin, Chance, Order, Change (Three Colours), 1982, 68 × 68 cm, screenprint in EXACTA, dal Construttivismo all'Arte Programmata (Milano: Fausta Squadriti Editore, 1985). Analysed by Jeffrey Steele, Chance, Change, Choice and Order *Leonardo* **24** 407 (1991).

Because the entropy concept is used in biological sciences, economics and political theory, as well as in statistics and cosmology, we may regard it as reminding us of the unity of knowledge. That is also why many great minds have struggled with the entropy concept: Einstein, Fermi, Nernst, Popper, Planck and Schrödinger, to mention only a few.

As a summary, entropy, when based on probabilities, gives one a *measure of spread* of the probability distribution. For a narrow distribution, when a few states have collected most of the probabilities, we have near certainty and the entropy is low. For a widely spread distribution we have much uncertainty and the entropy is high. Indeed, with any probability distribution one may associate an entropy value (by using an appropriate formula).

Here is an analogy. Imagine an isolated community in the Himalayas [4.3]. It has to pay porters to bring food and fuel up the mountain and adopts an unusual system whereby the 'happiness' of the community is always required to increase in the following sense: if a rich person is asked to pay out £100 for the transport of food and fuel, then a poor person is given £1 (say). The idea is that the loss in happiness of the rich is at least balanced by the gain in happiness of the poor. The difference (£99 in our case) is then available to pay for the transport of essentials up the mountain, the 'happiness' of the community being unchanged. Happiness? Well, I mean this rather narrowly defined financial kind of happiness which is needed for the story. We shall ignore all other activities in the community.

The money transactions described above correspond to heat exchanges. Since the money for the porters is the difference between what the rich man loses and the poor man gains, the total amount of money in circulation in the enlarged community which includes the porters is unchanged. This mirrors the law of energy conservation in physics. Similarly, money flowing from the rich to the poor corresponds to heat flowing from hot to cold bodies. Thus 'richness' in this community corresponds to 'hotness'. Lastly, our statistical concept of community happiness is the analogue of the enigmatic entropy. Thus our community's unusually altruistic constraint of increasing happiness illustrates an important property of the entropy: in heat exchanges the entropy of an isolated system cannot decrease. In fact we have with this conclusion reached the main content of the second law of thermodynamics (section 2.7, p 21).

The perceptive reader will realize that these money (or thermal) transactions cannot go on indefinitely in an isolated system. When everyone in our community is equally rich, the happiness constraint will prevent them from paying for any more services. Similarly, in thermal engines; typically you need to make steam to drive turbines, which in turn rotate machinery or perform some other type of mechanical work. We would be hard-pressed to produce any work if the whole system, steam, turbines and the rest, attained a uniform temperature! After all, the water has to be introduced into the boiler where it is heated, vaporized and eventually expands as steam against the turbine blades, condenses and returns as water to its initial state. As seen in section 2.7, a thermal engine requires at least two temperatures.

In addition, returning to our analogy, if a foreign conqueror starts to exact levies, the community's happiness will suffer—but the community is then no longer an isolated system, a condition which we had carefully imposed . How do we, in analogy with the levies extracted by the conqueror, cool our thermal system? We stop keeping it isolated and place it in a refrigerator, of course! It is not surprising that, if a system is *not* isolated, then *anything* can happen to it, and in a refrigerator the entropy of a system decreases (while it increases for the refrigerator). Thus it is wrong to say, as is sometimes done, that 'entropy must always increase'.

It is clear that, for our Himalayan community to have most money available for the porters, poor people should be given just enough money to prevent the overall decrease of happiness. In that sense our community is not all *that* altruistic! In our thermal systems the situation is exactly the same: the more entropy increases in a process, the less work can be produced by this process. Wastage arises if a rich man gives money directly to a poor man (without considering the porters): happiness goes up, but no work is facilitated. Similarly, if a hot body is put into contact with a cold body, a potentially work-producing temperature difference is uselessly dissipated. This loss of an opportunity corresponds, for example, to giving away your home instead of selling it!

We see that it is not sound science to ask people to avoid 'wasting energy'. Energy is normally conserved (section 2.5). The request should be not to generate unnecessary entropy!

Now let us consider an application. Many satellites are powered by solar cells which convert solar radiation directly into electricity. A measure of the 'efficiency' of the device is then the electric power produced as a fraction of the solar power input. Now the energy conservation law simply states that no energy must be lost in this transformation, and so it gives a maximum efficiency of unity. This exciting possibility is, however, wrecked by the second law which implies that heat must be given to a second body (remember the 'poor person' in our community!). Thus not all the incident energy can be used and the theoretical efficiency drops below unity. This corresponds again to the need for two reservoirs for the engine.

A very interesting property of the entropy is that it tends to increase as things are jumbled up, i.e. with disorder. We can see this in two steps. First, people become happier as they get richer; in our analogy this means that entropy tends to increase with temperature. Secondly, a solid is based on a *lattice* which is a grid of regularly spaced atoms, each vibrating about its own equilibrium position. Thus, as a solid is heated, some of its atoms are displaced more or less permanently (by thermal agitation) from their proper lattice positions— this is the beginning of disorder in the lattice. So we now appreciate that disorder goes up with temperature and hence with entropy. (In section 5.6 we shall have to modify this conclusion in the light of a more refined way of looking at this problem.)

If a system is *certainly* in one particular state, then it cannot possibly be in any of the other states. In this case of certainty as to the state of the system, the mathematical definition (which we do not need here) is such that the entropy is zero. Zero entropy therefore specifies the states of certainty. At the other extreme all the many states may be equally probable: the entropy has then the largest possible value for this system. The first situation is 'orderly' (figure 4.1!) the second is 'disorderly' (figure 4.2!). It turns out that for any given system the equilibrium value of the entropy is the largest possible which you can find by juggling about with the values of the probabilities. For an isolated system, then, equiprobability of its states means maximum entropy (table 5.1, p 112, gives an example).

Now we can discuss one of the big questions. Regarding the universe as an isolated system, and going back in time, we should reach a state of high orderliness. This was interpreted as due to an initial creative act, well before the Big Bang models of cosmology came into vogue. Some people even thought that here was a proof for the existence of God (Chapter 9)! Similarly, they pointed towards the eventual future state of uniform temperature in which no more work could be performed, and therefore life would cease to exist, and called it the heat death of the universe. These conclusions are somewhat modified, but not destroyed, by the notion of the expansion of the universe and other more modern ideas.

Ludwig Boltzmann (1844–1906) was probably the first to face the problem of why in a broadly static universe (as it was then assumed to be) the sun was still shining. If the universe is very old, as was assumed at the time, you would expect thermal equilibrium instead. He proposed as a way out that we are living in a giant fluctuation from equilibrium which is confined to our region of the universe. With the Big Bang models of cosmology, discussed in Chapter 7, this notion is now only of historical interest.

Just as scientists began to think that no more surprises were to be had from the entropy concept and the second law, the existence of black holes became an interesting speculation. A feature of black holes, discovered by theoreticians in the 1970s, is that you can associate with them a temperature and an entropy. The result is that you would expect normal black holes which are near each other to merge on purely thermodynamic grounds. Why? Because it turns out that merging would increase the entropy of the system!

The philosophy underlying this book bids us ask again: must this always be true—are there no exceptions? Indeed: in some cases the entropy increases if the system fragments into smaller systems [4.4]. Further, can the system be 'isolated', as mentioned above, for a macroscopic period of time? The answer is: 'strictly speaking, no'. We have here yet another idealization which can be approached, but not actually realized. The energy levels of a large system are so closely spaced that the movement of an electron within a few metres represents an outside disturbance that can cause a transition [4.5]. Further, on purely logical grounds, you cannot interact with a system which is strictly isolated!

With these cautions understood, we can begin to see how the second law leads to fascinating conceptual questions (and even to some answers!). And we have not even discussed it yet in relation to the theory of evolution, or the direction of time in a contracting universe. Nor have we touched on its colourful history! We will come to these questions in Chapter 7.

4.2 How can we move in time?

Suppose a billiard ball bounces off the side of a table at a point P, giving a trajectory APB. The laws of mechanics also allow the reversed trajectory BPA. Since this is the same as the original trajectory, but with time reversed, we can say that the laws of mechanics are time-symmetrical. A most important question concerns the boundary conditions. How was the ball projected: from A or from B? That decides in which of the two senses the ball follows its trajectory. Note, therefore, that for collisions among hard balls (or particles or atoms) the laws of mechanics which govern them are time-symmetrical. (We rule out the use of elastic balls as they would be subject to compression and heat would be dissipated and we are in this section interested in mechanical, not in thermodynamic, effects. But we are again adopting a serious idealization.)

Let me explain this *reversibility* differently. Suppose two atoms come along from two directions, collide, separate again, and move off in two new directions. This is called an elastic collision because it is just like the bounce of an elastic ball from the ground: it bounces back unaltered. Suppose next that the balls come in, in a second collision, from the final directions with their final speeds, collide and separate. How will they *now* emerge? They will have precisely the speeds and the directions they had *initially* in the first collision. It makes sense, therefore, to call the collision 'reversible'. Here is an alternative way of thinking about this. Suppose you had made a home movie of the first collision. On running the film backwards, you would see a second collision. Would the laws of physics allow such a collision with such speed and such directions to occur in real life? The answer is 'yes'— for this is precisely the second collision we talked about before. Let us try a third way. If we were able to run time backwards, then the first collision would yield the second collision. We shall call such processes 'T-invariant', which will be seen to be an important concept.

Thus we can take our pick in which of three ways we want to look at reversibility. First, the first and the second collision have the same directions and speeds; second, the normal and the reversed movie give real collisions; third, the time machine thrown into reverse gear also gives real collisions. This reversibility also holds for photons since you can see your friend's eyes by reflection in a mirror, and he can see yours: the light rays have reversible paths. Indeed: suppose a light illuminates a room. So it is conceivable that the walls could emit radiation, spectacularly well matched as regards their energy and direction, to light up the bulb in the middle of the room, while conserving energy! This is of course unlikely, but allowed in principle. Reversibility also holds for elementary particles generally—with one extraordinary exception. This is kaon decay. Its violation of time-reversal (i.e. of T-invariance) was inferred rather indirectly in 1964 (leading to a Physics Nobel prize for J W Cronin and V L Fitch in 1980). In the period 1995–1998 this violation was confirmed by direct measurements at CERN.

In principle, the unsolved problem of the forward march of time in physics could conceivably be explained in terms of particles by making use of this microscopic arrow of time! This view has had support recently in the study of short-lived B-mesons (consisting of an anti-bottom and a down quark). The violation of time reversal symmetry is here stronger than in kaon decay which is thus no longer an isolated curiosity, and a major advance in the understanding of irreversibility in terms of particles becomes a possibility.

In analogy with T-invariance, we have C-invariance if the system follows the same laws when particles are replaced by their antiparticles, and P-invariance if the same laws hold when the system is replaced by its mirror image. Scientists refer to C as 'conjugation' and to P as 'parity inversion'. All systems are believed to follow the same laws if the operations of T, C and P are applied jointly, i.e. they are 'TCP-invariant'. As an example, neutrinos υ and antineutrinos $\bar{\upsilon}$ have a spin about their direction of motion like a corkscrew, namely a left-handed and a right-handed one respectively. The C-operation makes Cυ a left-handed antineutrino which does not exist! CPυ gives a right-handed antineutrino, which does exist. Thus C and P are not valid symmetries in this case. C and P applied at the same time, i.e. CP, is a satisfactory symmetry—except that we know that the rare incidence of kaon decay actually violates it. This still leaves TCP as a satisfactory symmetry. In other words, a film of a particle event should give the correct physics of the antiparticle (C) if it is run backwards (T) and viewed in a mirror (P)!

Consider next a box of gas, the ends of which are at different temperatures. It will, by heat conduction, attain a uniform temperature after a little while (if otherwise isolated). How does this happen? By molecules colliding with each other, and the faster molecules from the hotter part losing energy to the slower molecules of the cooler part. But each collision is time-symmetrical, so that the reversed process, a kind of anti-heat conduction, is also possible, though this reversed process is never seen. The laws of mechanics, suggesting the sequence

temperature difference – collisions – uniform temperature,

should therefore be matched by the *anti-heat conduction*

uniform temperature – collisions – temperature difference.

We already know that entropy goes up for the process in the first line. It must therefore go down under anti-heat conduction; we can call this behaviour anti-thermodynamic. Our problem can therefore be expressed as follows: thermodynamics implies entropy increase; mechanics allows entropy to decrease; so the two subjects do not agree. Mechanics seems to be the loser as we 'never' see the anti-thermodynamic behaviour allowed by mechanics.

Again, if a gas is confined by a partition to one half of its container, then, upon removal of this partition, you have a non-equilibrium state, which gives rise to diffusion and thereafter to an equilibrium state in which the gas fills the volume uniformly. Time-symmetrical collisions are again responsible, and we ask: why is the reverse process, the spontaneous contraction of the gas into a part of its container, a kind of anti-diffusion, 'never' seen? This is another variant of the earlier problem.

Box 4.1 Maxwell's demon.

To make progress a short historical interlude is helpful, in order to introduce demons D2 and D3 from the list below. The list offers the reader the seven demons which will be noted in this book. Demons are here defined by the jobs which they can do; these are always beyond human capabilities.

D1 1812 Laplace: all-knowing (this chapter, p 86)
D2 1867 Maxwell: anti-heat conduction (this chapter, p 79)

D3 1869 Loschmidt: velocity reversal (this chapter, p 81)
D4 1936 Eddington: particle count for the universe (p 220)
D5 1937 Dirac: decrease of the gravitational 'constant' with time (p 219)
D6 1970 Landsberg: expansion or contraction of the universe (p 189)
D7 1987 Eigen: conversion of inanimate into animate matter (p 117)

(I apologise for D6 in this illustrious company, but this term was invented by Professor H S Robertson, University of Miami, during a 1970 conference [4.6]).

I would like to invite you now to travel, in your mind, to the Vienna of 1895. It was a vibrant city, with great intellectual talents. It produced many people who were to contribute to European life and culture. On the musical side there were Gustav Mahler, Johannes Brahms and Arnold Schönberg. The dramatic arts were represented by people like Arthur Schnitzler and Hugo von Hoffmannsthal. In addition, Sigmund Freud was pioneering a new science. Our interest, however, lies in the University of Vienna. This ancient foundation had attracted in 1894, to take the Chair of Theoretical Physics, the foremost proponent of the atomic theory (already encountered in section 3.5), in the form of Ludwig Boltzmann (1844–1906). He made it one of his main ambitions to give an explanation of the increase of entropy in isolated systems through the study of the motion of atoms. We shall therefore call this the Boltzmann problem. The respect of physicists for him is evidenced by the fact that a Boltzmann medal for work in statistical mechanics was instituted in about 1970. He also has a constant named after him, and he is our *hero* for this chapter.

Two particularly noteworthy people accompanied him on his journey of discovery. James Clerk Maxwell (1831–1879) was a great Scottish theoretical physicist who worked on thermodynamics and his 'Maxwell equations' govern the phenomena of electricity and magnetism. In comparison with this, the demon he invented is one of his minor claims to fame. The Maxwell house recently established in Edinburgh is further evidence of the high regard in which he is held. Secondly, Joseph Loschmidt (1821–1895) took, as another Austrian, a keen interest in Boltzmann's work and suggested an experiment which was regarded as impossible to perform in his days, but, as we shall see, can now be done.

In order to produce anti-heat conduction one can construct Maxwell's demon. Starting with an equilibrium gas, he opens a trap door in the gas, so as to let through the fast molecules to the left (say), but does not let the slow molecules through. The demon has thus separated the fast from the slow molecules.

We know already that any large or macroscopic system can also be viewed as a microscopic system—by focusing attention on the atoms or molecules for example. The temperature of a system can be interpreted as resulting from the motion of atoms or molecules: in that sense it is an *emergent* quantity; thus, who has ever heard of the temperature of a molecule? Now the faster the molecules, the hotter the gas. So the demon has established a temperature difference: one side of the box has become hotter, so that an *anti-thermodynamic* process has been performed [4.7, 4.8]. To bring about anti-diffusion one can similarly instruct Loschmidt's demon, who reverses all molecular velocities at a certain instant. The result is, of course, that all the molecules return to the original corner of the box. This is indeed anti-diffusion and it is 'never' seen!

The word 'never' is a very strong word, and it can easily lead to incorrect statements. Thus for a gas of two molecules it is certainly true that it will occasionally and spontaneously contract into a portion of its container. Even for five molecules this is true. It is because we implicitly assumed that the gas contains many, many molecules that there is a Boltzmann problem at all. We now see that the reason that there is a problem is connected with the statistics of particles. Even if the number of molecules is astronomically large, a gas in equilibrium will return to a microscopically defined initial state, but you may have to wait for a long time, which can be calculated. If that time exceeds the period from the last Big Bang to the present, i.e. the age of the universe, then we can safely dismiss the possibility of actually seeing such a return.

A card shuffling example is often given in this context (e.g. see [4.9]). We take a pack of 52 cards, which is a tiny number compared with the millions and millions of molecules in even only a cubic centimetre of a typical gas. The probability of regaining the original arrangement of the cards after a proper shuffle is nevertheless tiny: less than one in a million...million (the word 'million' should appear eleven times). If every person on earth were to shuffle a pack once every second, it would still take statistically much longer than the age of the universe to hit on the original arrangement in which we are interested. This gives an idea of the effect of the large numbers involved.

Because there are so many card sequences which are *not* the original arrangement, we say that the 'statistical weight' is heavily against us.

The distinguished scientists of the last century could not have foreseen the arrival of the computer. The relevance of computers here is that they, too, convert energy into rejected heat, while at the same time producing some mathematical results. There is energy dissipation here, and computers were regarded until the 1970s as *essentially* entropy-generating (i.e. 'irreversible') devices. It was then realized that the computing process itself could actually be made reversible (E Toffoli, C H Bennett, R Landauer), i.e. without entropy increase.

There *is* an irreversible step, but it lies not in computing itself, but in the clearing of the computer's memory after the computation, so as to prepare it for the *next* computation. The old Carnot cycle is actually not all that different. These cycles can, of course, also be carried out reversibly. But should you by any chance wish to restore the engine to its starting state (after, say, one cycle), you would have to supply to the hot body (or hot 'reservoir') the heat which it had lost in the cycle, and to tap off from the cold reservoir the heat it had gained during the cycle. These would indeed be highly irreversible operations, and would correspond to the wiping clean of the computer memory.

As a result Maxwell's demon has risen, phoenix like, to enliven the world of computer theorists. For example, double-cycle engines have been discussed (but not actually made, as far as I know) in which there is thermal coupling, as of old, together with new additional 'information-coupling' [4.10]. But again, as expected by now, the second law survives in its new generalized setting.

In the end, however, a *modern demon* makes observations and prints out the data about the molecules on a ribbon of initially blank paper. Using this information, the demon can draw energy from the system, violating the second law. But there is a snag. The ribbon is finite and has to be cleaned up from time to time, and when this operation is taken into account the second law is again saved. Suggestive and lively, the demon has not been killed off, but he has made us think again and again: that was Maxwell's purpose in creating him.

4.3   The first problem: can all molecular velocities be reversed?

The *first* problem (anti-diffusion) can be understood by accepting the fact that velocity reversal accomplished by Loschmidt's demon *does* decrease the entropy. Boltzmann knew, when he challenged Loschmidt: '*You* reverse all the molecular velocities', that it could not be done. But now *we* can do it—because of the master toy of the twentieth century: the fast computer. The collisions are simply traced backwards on the computer by using the calculated atomic trajectories. Indeed the entropy of the gas decreases according to these computer experiments! Thus initial conditions do exist from which entropy decreases (e.g. [4.11]). These initial conditions are a special type of boundary condition, as introduced in section 4.1, and are now *constructed* by velocity reversal. They are, however, so extraordinarily delicate that slight computer approximations bring you back to entropy increase. You may call this an *instability* of the boundary conditions.

The card experiment mentioned above gives us some idea of the vast number of possibilities generated by even quite restricted situations. This number becomes even larger when you consider various initial conditions for anti-diffusion. The arrangement we are looking for simply drowns in the sea of possible alternatives.

You can see that the Loschmidt velocity reversal was perfectly good mechanics, so that no purely dynamical proof (i.e. one based on classical mechanics) of the second law was possible. Next, the demon cannot violate the second law since Maxwell's original analysis neglected energy dissipation which enters via the reflection of photons from the molecules. This is essential if the molecules are to be 'seen' by the demon. It shows that the establishment of the second law by the methods of theoretical physics requires one to go beyond classical mechanics: the second law is not a law of mechanics and cannot be violated by experiments in mechanics (Szilard, Brillouin [4.12]).

People also began to realize that equilibration proceeds in a physical system because the number of available states for equilibrium is vastly greater than the number of states available for any given non-equilibrium state. So the second law of thermodynamics was recognised to be 'merely' a statistical law.

4.4   A second problem: coarse-graining

We do not here require a complete understanding of the notion of 'coarse-graining'. A rough picture will do, and in connection with table 4.1 (below) a fuller appreciation of this concept will emerge.

We have treated anti-diffusion, anti-heat conduction and therefore the difficulties associated with the demons of Maxwell and Loschmidt, as the *first* problem of irreversibility. The *second* problem can be formulated by taking all the classical and quantum mechanical information about the system into account. It is then found mathematically that the entropy of an isolated system cannot change with time. But we can also see this intuitively: it is precisely the time reversibility of classical and quantum mechanics, discussed in section 4.2, which suggests that any system can move forwards and backwards in time without any problem. So entropy cannot increase in one direction, for it would decrease in the opposite direction, destroying the presumed time symmetry. The only remaining possibility is that the entropy remains constant in defiance of the second law, which requires entropy to increase *provided* we are dealing with a non-equilibrium system. For entropy *increases* as it attains equilibrium. Our second problem is to extract this increase from statistical mechanics. In spite of heroic attempts by Prigogine, Nicolis and their Brussels school, the Dutch school of L van Hove and N van Kampen, and by many others, this problem has not been resolved to everybody's satisfaction.

(Technically the system of N particles may be represented in an imagined many-dimensional space in which all $3N$ position coordinates and all $3N$ momentum components are the axes (three of each per particle, see p 18 for the meaning of momentum). The *state* of the whole system is represented by a single point in this imagined space, as explained in section 4.1, p 68. This space is called a phase space. A point moves in this phase space as the state of the system changes according to classical or quantum mechanics, which both yield time-symmetrical ('T-invariant', see p 75) equations of motion. If we now consider many copies of the same system, each started off differently in its motion, these are then represented by many points in phase space, giving rise to 'phase space densities' of these points. It turns out that the non-equilibrium entropy, which is defined in terms of such phase space densities, is *constant* in time whereas thermodynamics requires it to increase (or remain constant!). This is precisely what was explained in different language in the preceding paragraph.)

Let us look for a way out of this difficulty. If all the available classical or quantum mechanical information has been used to describe a system, we can speak of a fine-grained entropy. We have noted that the fine-grained entropy is normally constant in time. Hence, by a rule of logic (see below), an increasing entropy implies that it has to be non-fine-grained. The term used is *coarse-grained*. It means that we confine attention to *groups* of classical or quantum mechanical states, and say whether or not our system is one or other of these *groups*. Under these conditions entropy can indeed increase with time, though this is purchased at the expense of a loss of detailed knowledge about the system. But for thermodynamics such detailed knowledge is not needed. In any case we again have to accept an incompleteness in our knowledge which dominates this book, and is signalled in this case by the the need to use probabilities.

The 'rule of logic' used above (see also figure 4.3) is very simple. It runs as follows:

'If all swans (i.e. S) are white (i.e. W), or briefly S implies W, then a black object (i.e. not W) cannot be a swan (not S).'

The rule is therefore clear:

'If S implies W, then not W implies not S.'

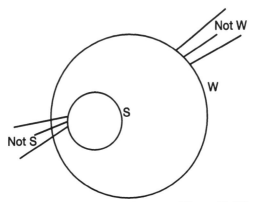

Figure 4.3 Illustration of a rule of logic (in S → in W; outside W → outside S).

Thus when a gas expands into a neighbouring, previously empty, container its entropy goes up because, unable to give a full atomic description, one uses a coarser, more thermodynamic, account. A full description is in fact never possible because of the many uncontrollable influences which act on the system (see the end of section 4.1). It has been pointed out [4.13] that some well-known physicists seem to have made erroneous statements of this matter, and some do not approve of coarse-graining.

I now give you an example of coarse-graining—just to illustrate the procedure, not to derive any special results. Four boys of equal mass play on a see-saw, and we shall write (ABC,D) if A, B and C are on the left, while D is on the right. Springs stop the see-saw from touching the ground. There are 16 possible 'microstates' shown in table 4.1. But suppose we are too far away to distinguish the boys, while we can clearly see the angle of the see-saw. There then result the five possible macrostates given in the table. Our short-sightedness leads to coarse-graining (for us)!

Table 4.1 Micro- and macrostates for boys A, B, C and D of equal mass on a see-saw. The 16 microstates are arranged in terms of the five numbered macrostates.

1.	ABCD,-
2.	ABC,D; ABD,C; ACD,B; BCD,A
3.	AB,CD; AC,BD; AD,BC;BC,AD; BD,AC; CD,AB
4.	A,BCD; B,ACD; C,ABD; D,ABC
5.	-, ABCD

In macrostate 3, which consists of the six microstates shown, the see-saw is perfectly balanced. In macrostate 1 it is inclined strongly in one direction; in macrostate 2 it is less strongly inclined in the same direction. In states 4 and 5 it is inclined in the opposite direction. The probabilities of the macrostates can be seen to be 1/16, 4/16, 6/16, 4/16, 1/16 respectively. This leads to a smaller entropy for the 'macroscopic' description than for the 'microscopic' description, because the number of states involved in the former is smaller. Now we have seen that the expansion of a gas into a vacuum by the removal of a partition increases the entropy. We find the same feature here. By constraining the boys to positions which keep the left-hand portion of the see-saw down, we reduce the number of microstates from 16 to five, and this can be shown to reduce the entropy. Conversely, removal of this constraint causes an increase of the entropy.

The procedure of coarse-graining is not accepted by all workers in the field, and, indeed, it is not always appropriate: suppose a viscous fluid is contained between two concentric glass cylinders, and a drop of insoluble ink is placed in the fluid. Upon slowly turning the outer cylinder the droplet is drawn out into a thin thread and eventually becomes invisible. A coarse-grained description could not account for the fact that upon reversing the motion, very slowly the thread goes on to reconstitute more or less the original drop. In comparatively rare cases such as these, when the original order is not lost, although it appears to be lost, coarse-graining can mislead. This type of problem was discussed by Gibbs [4.14] and was used as an example of 'enfolding' of information [4.15]. Many other modern experiments make use of similar ideas [4.16].

General guidelines of how to design macrostates in terms of microstates still need to be developed, and they would have to tell us what size the macrostates should be. If this is achieved in the future, the apparently subjective feature involved in the design of particular coarse-grainings may disappear. But it could well be that such a theory would have to be so general as to be impossible. It has in fact recently been remarked that 'the most notorious unsolved mystery of statistical mechanics is...the problem of irreversibility' [4.17]; for example, an extended effort since 1973 in that direction has not convinced the scientific community that a solution has been obtained [4.18].

Of course if the system of interest cannot be safely regarded as 'isolated', new mechanisms for the increase of entropy with time become available. Notable among them is the interaction of the system with the environment. Correlations among the particles in the system are then reduced and coherence (see p 146 for a brief discussion) is gradually lost. This loss leads to an increase of entropy with time, just as coarse-graining does.

4.5   Time's arrow as an illusion

The increase of entropy provides one of the key physical indicators of the progress of time. If it can be extracted from mechanics only by coarse-graining, the question must be faced of whether the progress of time itself is only an illusion. To consider this matter, let me call in Laplace's demon, often called the Laplacian calculator. He is our oldest demon, hails from 1812, and I have called him D1 (see p 77). He is a dematerialized intelligence, a kind of God who knows of all collisions, can distinguish all microstates in a fine-grained phase space, and all his calculations of future and past states (in so far as allowed by science) are performed instantaneously.

For him all elementary processes are therefore time-symmetrical (the time-symmetry violation by kaon and similar decays apart). As he knows only of these elementary processes, how would we communicate with Laplace's demon? Maybe time would not exist for him. The concept 'table', for example, is far too rough for him: where we see a surface, he sees a swarm of molecules which mix with those of the surrounding air. We would have to tell him to discard information which he has at his disposal, so as teach him our language, which uses words for our rough concepts; and again arrive at coarse-graining so as to extract the direction of time. This recalls an often-quoted remark of Einstein's to his close friend Michele Besso, who kept enquiring about the nature of irreversibility. Einstein considered it to be an illusion produced by improbable initial conditions. Further, on Michele's death Einstein wrote to his son and his widow that 'For us convinced physicists the distiction between past, present and future is an illusion...' [4.19].

These considerations are in the Greek tradition: Zeno of Elea attempted to show that motion was impossible by his well-known paradoxes (see section 6.1), but this did not fool the 'ordinary' person. Similarly the latter orders their life on the assumption that past, present and future have true physical significance, in spite of what some physicists may tell them. Everyday life is normally not in need of the more profound truths!

When I wrote this section I dreamt that I encountered some super-human intelligence in the guise of a shadowy demon. He turned on me in rage and thundered: 'You, little man, average or coarse-grain because you are ignorant. I have a thousand eyes and a million brains and I can take in at once which molecules are doing what. By coarse-graining you remove the individuality of these arrangements which I can distinguish, even if you cannot. You', he added scathingly, 'have to do this to produce the idea of a flow of time and thus to create a crutch for your weak little brain—but you have achieved nothing.' I was downcast in my dream, but replied in as dignified a way as possible: 'I realize that coarse-graining, which extracts a direction of time for me, is not needed by you'. After some reflection I pressed on: 'In fact, for anyone with all your knowledge, the universe must be stretched out in front of them to be comprehended, including its history and its future, in all its detail, in a single leap of their powerful mind. Perhaps they do not even require the notion of time?' It was a question, but the demon did not reply. He had disappeared (rather conventionally) in a puff of smoke.

The modern debate on the philosophy of time goes back to a paper of 1908 by the Cambridge philosopher J M E McTaggart who distinguished a tensed theory of time (his A series) from an untensed theory (his B series). The former requires temporal descriptions, e.g. 'the baby was born yesterday', which become faulty as time goes on. The latter utilizes descriptions which remain valid, e.g. 'the baby was born on 1 January 1997'. This distinction has given rise to a great deal of discussion among philosophers. In science we tend to use the B series, but do so implicitly. McTaggart or the B series are usually not mentioned [4.20]. There are other ingenious ideas about, for example some seek to treat past and future even-handedly [4.21], while others deny the existence of time altogether.

4.6   Different arrows of time

The 19th century already knew several arrows of time: (i) the psychological one (we remember the past, not the future), (ii) the biological arrow (evolution), (iii) the thermodynamic one (entropy increase) and (iv) the electromagnetic one (radiation comes from a light bulb, rather than converging onto a light bulb to light it up), which came a little later. This century brought us (v) the arrow of kaon decay of subnuclear physics (see p 76), and (vi) the cosmological arrow of time which gives us at present an expanding rather than a contracting universe. Why do these arrows all point in the same direction? Although it has been suggested that the cosmological arrow is primary and impresses its direction on all the others, this is not generally accepted: we are clearly reluctant to believe that milk and coffee get mixed up in our cup because of the expansion of the universe. In other words, it strains our credulity to attribute the entropy increase due to the mixing of fluids in one's cup to the expansion of the universe!

Less controversial is the view that, since memory traces laid down in the brain are biochemical, this arrow determines that (i) the biological, (ii) the psychological and (iii) the thermodynamic arrows all point in the same direction. You could go one step further and blame the 'accident' of our physiological make-up for all the problems we have regarding time, its meaning and its asymmetry. A dematerialized intelligence could clearly grasp all developments in a flash, and without a time coordinate (as we saw in the preceding section). That would indeed be a solution of the 'problem of time'.

Alternatively, just as temperature is a quantity which can be regarded as emerging from the molecular or atomic picture of a system (a single molecule has no temperature), could the same idea be applied to time? To say so would be to affirm an act of faith. No final decision can be made, but a NATO Advanced Workshop took place in September 1991 when 42 clever physicists were clearly having fun down in the province of Huelva, Spain, from which there emerged 35 papers [4.22]. A poll was conducted (by Julian Barbour) on the question:

Do you believe time is a truly basic concept that must appear in the foundation of any theory of the world, or is it an effective concept that can be derived from more primitive notions in the same way that a notion of temperature can be recovered in statistical mechanics?

Of the 42 persons questioned, 10 believed that time exists at the most basic level, while 20 denied this and 12 were undecided. Here then is another question to which we have not yet an agreed answer, and incompleteness reigns.

Box 4.2 Riding on a photon.

It is widely known that if space travellers move at a high speed relative to the earth, then upon their return they are found to have aged less than their earth-bound friends (see p 202). Indeed, if they approach the speed of light very closely, then their clocks can be regarded as having been arrested almost completely, and they could visit us again and again, almost untouched by the ravages of time. To such travellers everything would happen almost at once. Indeed a free packet of light energy, a photon, could, by reflection between mirrors, visit the same place again and again in the same instant of *its own time*. In this sense it is *non-local*: it cannot be associated with a given spot within the volume available to it. Gravitational interaction is similarly non-local, since classically it acts instantaneously everywhere. As a model of our disembodied and all-knowing intelligence, we could think of it as riding on a photon. I want to leave you with the thought that it might not need the notion of time. Here, in the virtual disappearance of time, Christopher Marlowe's Faustus would find hope at last:

> *Stand still, you ever-moving spheres of heaven,*
> *That time may cease and midnight never come;*
> *Fair Nature's eye, rise, rise again, and make*
> *Perpetual day; or let this hour be but*
> *A year, a month, a week, a natural day,*
> *That Faustus may repent and save his soul!*
> *O lente, lente currite, noctis equi!*

4.7 Entropy as metaphor

The unity of knowledge is a doctrine which says that 'intellectually close' to any fact or any argument there are other facts or arguments which are also important, and that these intellectually neighbouring facts or arguments are illuminated and rendered more easily under-standable by the original fact or argument with which we started. It asserts furthermore that these relationships are reciprocal, so that the original fact or argument gains in stature and importance from these neighbouring facts or arguments. You can therefore imagine travelling through the whole field of knowledge by taking small steps from neighbourhood to neighbourhood until the whole of contem-porary knowledge has been covered. That no single person has the capacity to do so is a minor matter; the major matter is that a suf-ficient number of persons, when put together, can approximately simulate such an intellectual trip, or at least attempt a more restricted trip, which is beyond each person's own specialities. They will have to adopt among their principles the great guidelines of past philos-ophers: freedom from prejudice, love of truth, respect for others. The unity of knowledge thus becomes part of a philosophy of life, a guide even for those who do not see the abstract possibility of the 'big' trip and are happy to confine themselves to a 'small' one.

The entropy concept fits into this scheme not only because of its wide use in science, but also because it is a metaphor in the humanities. It is thus a vehicle which enables us to appreciate the unity of knowledge. Suggestive? Yes; but only a metaphor. It cannot be expected to prod-uce actual advances.

A bridge from science to literature may be accomplished via science fiction. As you might expect from metaphors, the use of the entropy concept here is only suggestive and often superficial. Its normal use is as a hint at ultimate chaos and at a general tendency to decay, be it of objects, people or social organizations [4.23]. M John Harrison talks of 'intimations of entropy everywhere' [4.24], and Colin Greenland talks of certain poems as containing 'messages of doom, disaster and entropy' [4.25]. The study of history has also not been immune from the incursion of entropy through Henry Adams' remarks on 'The Degradation of the Democratic Dogma' and 'The Rule of Phase Applied to History'. But they are now regarded by historians as rather esoteric.

Writers on art have also used the entropy concept [4.26]. More than sixty years have now elapsed since the famous mathematician Birk-hoff attempted to develop a numerical measure of beauty in terms of the ratio of 'complexity' to 'order' [4.27]. But his approach has not attracted many followers, though it has focussed attention on some interesting questions. For example, what interpretation can be given to 'order' in a work of art? In physics 'order' can be conceived by way of comparing a system with the system as it would be if maximally broken up and disordered by some standard procedure. Depending on the number of pigeonholes capable of receiving the component

Figure 4.4 Hans Baldung, The Ages and Death. Copyright Museo del Prado, Madrid: all rights reserved.

parts, you could then possibly use the entropy concept to arrive at a measure of disorder. It would depend on the number of such boxes, and in an artistic context this is not usually available. This metaphor may therefore never become useful. (A more systematic discussion of 'order' will be given in the next chapter.)

The sadness of the passage of time has often been commented upon in drawings or paintings (Hogarth) and in many religious contexts (figure 4.4).

An application of thermodynamics to economics [4.28], although initially favourably received, has lately run into criticism [4.29]. The basic concept is clear enough: the industrial society uses up resources in the process of production and so is expected to cause vast entropy increases. It has now been suggested, however, that this increase is actually rather small when compared to the entropy increase due to solar energy when it is received at the earth's surface without useful conversion (private communication from B Mansson, Engineering Science, Karlstad University, Sweden).

Last of all, is it really a logical necessity for time to be one-dimensional, as is argued by some philosophers? Think how you could circumvent unpleasant events by wriggling past them in a second dimension of time [4.30, 4.31]! I leave it to the reader to imagine, or even to write, an essay on the 'side-effects' of two-dimensional time. There is of course also Jorge Louis Borges, who in one of his stories envisages the realization of all possible outcomes of a situation. Here, contrary to the proverb, you can actually have your cake and eat it:

> *'Differing from Newton and Schopenhauer, your ancestor did not think of time as absolute and uniform. He believed in an infinite series of times, in a dizzily growing, ever spreading network of diverging, converging and parallel times. The web of time...embraces every possibility.'* [4.32].

Sixteen years later, in the hands of Hugh Everett III, this idea was reborn. It took the form of a cosmological model!

4.8 Summary

In Chapter 2 the laws of thermodynamics have been discussed, using the energy concept, but without the notion of entropy. Entropy is a hard idea, more difficult even than energy, for it is not obvious in everyday life: it is one of the concepts invented by scientists to help us understand nature. Although any formulation of the second law of thermodynamics requires us to introduce the idea of an isolated system, we have seen here that, strictly speaking, isolation can only be approximate—it does not really exist (section 4.1).

The understanding of the nature of 'time' has been a problem for centuries and, since the entropy of an isolated system tends to increase with time, entropy has often been regarded as furnishing a scientific key to the understanding of time. It tells us in what direction time increases. Sir Arthur Eddington referred to it as an 'arrow' of time [1.1]. I explained here the time-reversibility of mechanics, showed how it causes problems and how they may be overcome (sections 4.2, 4.3). This can be achieved by accepting, and making do with, incomplete knowledge through the introduction of statistics. In this way we can come to terms with the subject of 'statistical mechanics' (section 4.4). In conclusion, the possibility of 'time' as a human illusion was envisaged (section 4.5) and different arrows of time were distinguished (section 4.6). The notion of entropy is such a widely used metaphor (section 4.7) that it reminds us of a great vision: the unity of all knowledge.

As indicated at the end of Chapter 3, another area involving both microscopic and macroscopic ideas deals with chaos and life, to which we turn next.

Chapter 5

How a butterfly caused a tornado
Chaos and life:
science as synthesis

5.1   Introduction

How can life be extracted from dead matter? Following the Bible, let us start with chaos! The characteristics of chaos are relevant to a main theme of this book since they show that some expected information is not actually available. In the case of eclipses one can make very accurate predictions. They are as exact as required. Of course, they are not absolutely exact: there is always unavoidable experimental error; or fluctuations; or neglected gravitational effects from, for example, large comets. But *chaos* in the technical sense is something else again. For example, a system may be completely defined mathematically: the equations governing it are known; and yet we may not be able to predict its precise future. This can hold true even for some simple systems when they exhibit chaos. Of course the *approximate* future can often be predicted much more easily.

In a certain sense chaotic systems are often close relatives of systems which are predictable (although never completely so). It can be just a question of changing a parameter by a tiny amount which decides whether a system is chaotic or not. In that sense chaotic and predictable behaviour can be quite close, and a theory which covers both types represents a kind of synthesis. These ideas apply not only in mechanics, but also in the realm of chemistry and biology.

We also find in open systems (see p 22) that, as they depart further and further from equilibrium, they may suddenly snap into a new and more organized state, rather than showing chaos. The theory of these apparently distinct situations—chaos versus organization—reveals a closer relationship than might have been expected.

The phenomenon of self-organization is also important for living systems. Such systems can grow and so the number of quantum states available to them also increases with time. Indeed, we shall learn in section 5.6 that entropy and order can both grow at the same time. This is contrary to the usual view that, if entropy grows, so does disorder.

5.2 Limits of predictability in Newtonian mechanics

Something is always moving! On p 12 we noted that atoms keep moving even at low temperatures, and on p 39 Brownian motion was noted. Before Einstein's 1904 explanation of the effect, the well-known physicist and philosopher Ernst Mach did not believe in atoms. To anyone using this term, he would say 'Have you ever *seen* an atom?' Later he was converted by the evidence of Brownian motion.

Even at the lowest temperatures, when most substances are solids and their atoms are bound to lattice points (see p 73), there remain zero-point fluctuations of these atoms. It is in fact impossible to eliminate completely from experimental measurements the source of inaccuracy due to these motions. They represent a *first* irreducible limit to predictability.

This is usually not troublesome when human-scale type measurements are made, for the Brownian fluctuations are then negligible. That is why classical mechanics, i.e. the mechanics of Newton, is still being used to predict the motion of bodies.

A *second* limit is more sophisticated, and is connected with what has recently gripped public imagination under the general title of 'chaos'. You can easily appreciate it by observing an 'executive toy' showing the quite irregular and apparently unceasing motion of a magnetic pendulum as it moves over some little magnets.

The next few paragraphs will be devoted to explaining how chaos can occur.

When a pendulum swings under gravity we normally think of it as simply going forwards and backwards. If it can swing in *any* direction, so that its bob can trace out the shape of an inverted umbrella, it is called a *spherical pendulum*. Let us look at small oscillations (large ones would complicate the problem). A one-swing oscillation takes a certain time, T say. This is the *periodic time*. Now an unusual thing, which will produce exciting and unexpected results, occurs if another oscillation is forced upon the system, namely a forward and backward movement of the point of suspension of the pendulum. Suppose that this has another periodic time, T' say. In order to define this motion completely, the maximum displacement of the point of suspension has also to be specified (for the numerical results, given later, it will be taken to be one sixty fourth of the length of the pendulum, using for convenience the original papers cited). Suppose the initial displacement of the bob is exactly in the vertical plane given by the forcing motion. The pendulum will then swing happily in that plane.

The bob can swing in other planes as well. The reason is that the initial position of the bob can be known only within a certain error. This leaves a possibility for the initial position to be slightly out of the plane of the forcing movement.

If you start off movements with slightly different initial positions it is found [5.1] that the resulting motions will eventually diverge markedly from each other. This may be called the 'predictability horizon'. In our case [5.2] the forcing period T' has to lie between 0.989 T and 1.00234 T for this complication to develop. This is precisely an example of chaotic behaviour. Thus, as a result of a small initial inaccuracy, the position of the weight becomes impossible to forecast in the chaotic regime [5.1–5.3], as has also been checked experimentally [5.4]. An increase in the accuracy of the initial position from four to six decimal places provides only a modest improvement of the pre-dictability horizon. We are again in a realm of limited knowledge.

The last step towards the realization that even Newtonian mechanics is sometimes unable to lead to precise results is to note that Brownian motion, for example, limits the accuracy of the normally available initial data. But of course much more important limits may apply in any given case. You have a kind of synthesis between knowledge given by the laws of mechanics, and limited knowledge of the initial conditions. The specification of what happens at the beginning (or at the end) of a problem is called, more generally, the *boundary con-ditions*, a term often used in more specialized works.

The explanation of the need for approximate equality of T and T' to produce these effects depends on the notion of resonance. This is well known from a famous design problem which requires the natural period of oscillation of a structure not to to be too close to that of the disturbance. For example, soldiers marching across a bridge are advised to break step in case the periodic time (or frequency) of the march comes close to the natural periodic time of vibration of the bridge, causing major oscillations of the bridge and perhaps its result-ing collapse.

Chaos for a system implies that two closely similar initial conditions lead, after sufficient time, to widely different final conditions. That is why certain predictions are impossible. Chaos can readily occur in classical mechanics, but also in the study of fluids: turbulence of fluids is a well-known example of chaos; in the case of weather forecasting the fluid involved is the atmosphere. In formulating equations for weather forecasting the MIT meteorologist Edward Lorenz found in 1963 that even his relatively simple equations can lead to unpredict-able results. This was an important step in the history of chaos. He published his results in the appropriate *Journal of the Atmospheric Sciences*, which, however, is not read widely outside the meteorologi-cal community. That was 36 years ago, and this matter is still in the news. That is quite an achievement.

His result was rediscovered only in about 1970. Duly famous now, Lorenz was asked to address the American Association for the Advancement of Science in 1972. His title was [5.5] 'Predictability: does the flap of a butterfly's wings in Brazil set off a tornado in Texas?' This led to the notion of the 'butterfly effect'. This idea is of course problematical, as damping over long distances is bound to limit the range of the effect.

The interpretation of turbulence as involving chaos was suggested in 1971 by the Belgian-born physicist David Ruelle and by Floris Takens of Groningen. Thus a slow flow of water from a tap comes in a fine simple flow, but you can see turbulence when the water comes out chaotically on turning the tap on fully.

Chaos also occurs in many other systems whose laws are clear. But the initial conditions are subject to uncertainty. This is often called *deterministic chaos*. You learn that deterministic equations do not imply predictability! Alternative definitions of this and related concepts exist in the literature [5.6].

There are two ways of looking at the results from chaos, positive and negative. The positive way is as follows. Given the system and a time in the future for which a prediction is to be made, there will always exist some (possibly high) accuracy for the initial conditions that will achieve it. The negative way is: given a system and an accuracy with which the initial conditions are known, there will always exist a (possibly distant) time in the future for which a prediction cannot be made. A good way of looking at this antinomy is to regard science in the presence of chaos as a *synthesis* between predictability and lack of predictability: each has its own domain in any given case. This synthesis will be explored further in the following sections.

5.3   Chemical and population chaos

In figure 5.1(a) we see a 'tree' which develops branches each year. Each branch can grow a new one (drawn to the right and numbered) after it has existed for two years. By rearranging the branches somewhat, we can obtain a more convincing looking tree (figure 5.1(b)). This figure shows that the number of branches increases year by year in a simple way: 1, 2, 3, 5, 8,.... We have here repeated bifurcation phenomena, and the resulting series of numbers has the interesting property that each number is the sum of the preceding two numbers. This is a mathematically pretty phenomenon, and it is called the 'Fibonacci series' after the mathematician Leonardo of Pisa, son of Bonaccio ('Filius Bonacci' 1170–1250). Numbers from the series occur frequently in biological contexts. For example the number of petals of common flowers are often Fibonacci numbers: iris 3, primrose 5, ragwort 13, daisy 34, michaelmas daisy 55 and 89. They also occur in the study of arrangements of leaves on stems of plants, and the subject of phyllotaxis is well studied [5.7]. There is even a suggestion that the Fibonacci numbers played a part in Bela Bartok's compositions [5.8].

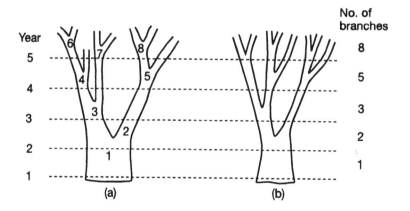

Figure 5.1 (a) A tree branching annually. A branch develops a new one after two years. (b) The same tree with branches rearranged.

Consider an open chemical system (p 22). Here the word 'system' can stand for many different objects, such as a pendulum, a box of gas, a boiling liquid, etc. Such systems are also usually in a container which may then be considered as part of the system.

Chaos can occur in these systems. Chemicals are fed into a chemical 'reactor', which is simply a container for the chemical reactants, where they are automatically stirred and the products of the reaction are drawn off. The technical means of achieving this need not concern us: for example the products could be arranged as solid deposits, and then removed. One particular chemical is arranged to be poured in as fast as it reacts, so that its concentration in the reactor remains constant. This is *not* an equilibrium state, since chemicals are going in and others are going out, so that there are changes occurring with time. But as the concentration of the chemical remains constant, the system is in a *steady state*. Another example of a steady state is provided by a bucket with a hole in it. It is in a steady state if the water stays at the same level while water is pouring in at the same rate as it escapes. A steady state is in fact the next simplest thing to an equilibrium state, and is therefore much discussed. In fact, if *all* concentrations are constant, the steady state can become an equilibrium state. We thus arrive at a *first* new distinction: between equilibrium and steady states. An equilibrium state is always a steady state, but not conversely.

We can depart further and further from an equilibrium state by speeding up the inflow and outflow of chemicals, as indicated by the horizontal coordinate in figure 5.2. This speed can serve as a measure of the departure from equilibrium of the system. A second new distinction to be made is that between *stable* and *metastable* (equilibrium or steady) states. A system is in a stable state if it returns to that state after a small disturbance has been applied for a short time. For example, a pendulum hanging under gravity returns to its position after having been given a small push. That is stability. But a book lying at the edge of a table can fall off after a small push: its position is not stable, and is called metastable.

Figure 5.2 shows the concentration of a chemical C plotted against the departure from equilibrium, measured by the rate of withdrawal of chemicals. At a certain stage the system suddenly has a choice between two alternative stable steady states, while the original state continues, but as a *metastable* state (dotted). This is called a 'bifurcation', labelled B, at which new branches develop on our curve. It is like a tree forming a new shoot. The equilibrium point is labelled E. Taken further, more bifurcations can crowd in upon each other and, depending on the nature of the chemical system, strange and striking results can be found. For example, the reaction vessel may change colour regularly, leading to a 'chemical clock'. Alternatively, many beautiful spiral patterns can result (e.g. in the Belousov–Zhabotinsky reaction, pioneered in Russia in the 1950s). If you proceed further from equilibrium these patterns disappear and the visual pattern which results is without structure: you then have essentially chemical chaos. Figure 5.3 (from [5.9]) shows how you can pass from an ordered state to a chaotic state and back again by appropriate manipulation of external conditions.

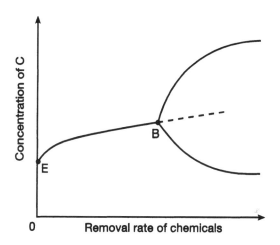

Figure 5.2 Bifurcation diagram for a chemical reaction.

We can also encounter chaos in the study of population growth. Let us start very simply. Suppose, for argument's sake, that the population of a country is always multiplied by the constant factor $r = 1.1$ to obtain the population $p(n)$ for the next generation, numbered n, say. This is shown in figure 5.4 up to $n = 3$. Then after 50 generations (not shown) you find an increase by the very large factor of 117. But if the multiplying factor r is less than one, say $r = 0.9$, the population will in due course die out (not shown). These phenomena are related to the Malthusian population explosion. It was Robert Thomas Malthus (1766–1834), economist and churchman, who pointed out almost exactly 200 years ago that poverty is inevitable if the population increases faster than the food supply.

In fact populations of animals and humans grow more slowly and erratically because of food shortage, illness, wars, etc. Hence a more realistic approach is to use a so-called *logistic curve* (figure 5.5). Here,

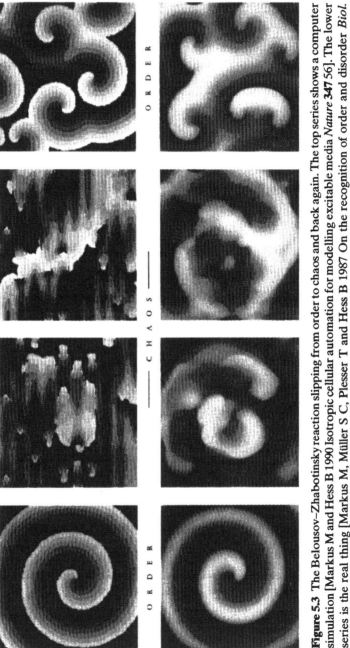

Figure 5.3 The Belousov–Zhabotinsky reaction slipping from order to chaos and back again. The top series shows a computer simulation [Markus M and Hess B 1990 Isotropic cellular automation for modelling excitable media *Nature* **347** 56]. The lower series is the real thing [Markus M, Müller S C, Plesser T and Hess B 1987 On the recognition of order and disorder *Biol. Cybernetics* **57** 187].

instead of passing to the population for a new generation by multiplying the old population simply by a factor (1.1 and 0.9 were used in the above examples), there is now an additional term which tends to reduce the population of the next generation. The equation is now 'non-linear' since it contains a term involving $p(n)$ multiplied by itself. It is these non-linear equations that have very interesting properties and will be seen below to lead to self-organization and chaos.

The curve for the $r = 1.1$ case is just the straight line of figure 5.4. Looking at it more carefully, we can imagine that we start with a population denoted by A in the diagram, and pass to populations B, C...for generation 2, 3, etc. It will be useful for later to develop this straight line in several steps. Having got to B, read off the population on the vertical axis, and draw a horizontal line to meet the vertical axis for $n = 2$. Now, for the second generation we have to multiply the B population by 1.1 to reach C. This is shown clearly by the dashed lines. Next, we get to population D for $n = 3$ by similar auxiliary lines, which are dotted this time for easy recognition. The result of this construction is of course the same as before, but it helps when we

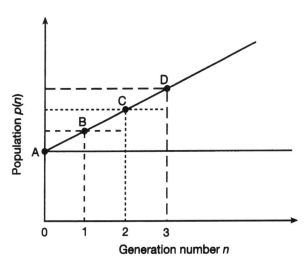

Figure 5.4 Increase of population with generation number.

come to figure 5.5. In that case, because of the non-linearity, the straight line ABCD... is replaced by a curve and the figure shows four examples.

Let me now write $x(n)$, a pure number lying between zero and unity, as a measure of the population for the nth generation. Figure 5.5(a) explains how to pass from the first generation with a population given by $x(0)$, via the curve, to the next generation with population $x(1)$, marked on the *vertical* axis. Next, starting with the value of $x(1)$ on the *horizontal* axis, one finds $x(2)$ on the vertical axis, and so on. The line $y = x$ is there merely to enable one to pass easily from $x(1)$ on the vertical axis to $x(1)$ on the horizontal axis, and similarly for $x(2)$. This is shown in figure 5.5(a), as it was in figure 5.4.

Suppose next that the parameter of the logistic equation (which I shall again call r) is such that the line $y = x$ (which is now dashed) lies *above* the curve (figure 5.5(b)). Then, starting from any value of x, we are led to smaller and smaller values of x. In figure 5.5(b), r has been chosen at 0.71 and we arrive at a final steady state in which the population has died out ($x = 0$). In fact, the population must die out whenever r is less than one, as we might have guessed from the original case ($r = 0.9$) we considered.

If $r = 2$ we find (figure 5.5(c)) a normal population with a final value of x given by 0.5.

Entirely new situations arise if r lies between 3.0 and 3.5. The population is now found to oscillate between two values. For example, for $r = 3.2$ (figure 5.5(d)) we do not get a unique answer. (The resulting situation is called a limit cycle). For values of r above 3.58, bifurcations are heaped upon each other and we find chaos, as shown in figure 5.6. This curve is often named after the American physicist Mitchell Feigenbaum who discovered many of these unusual properties of the logistic equation, which is a frequently cited example in the context of chaos and bifurcation. It exhibits a cascade of bifurcations which lead to chaos.

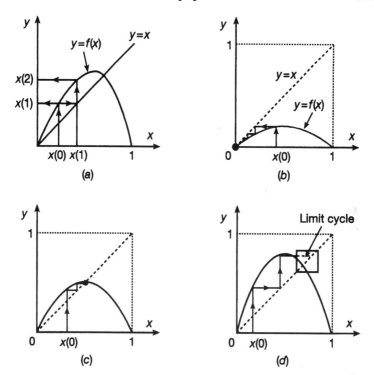

Figure 5.5 Final values of x as obtained by the logistic equation. (*a*) Procedure for obtaining $x(n+1) = rx(n)[1-x(n)]$, i.e. $x(1)$ from $x(0)$, $x(2)$ from $x(1)$, etc. (*b*) $r = 0.71$. (*c*) $r = 2.0$. (*d*) $r = 3.2$.

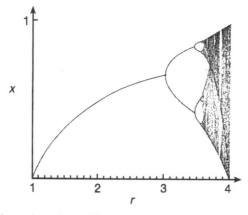

Figure 5.6 The path to chaos. The final values of x for various values of the parameter r of the logistic equation.

We see that a transition from definite final populations to a bifurcation (or a 'two-cycle') occurs for $r = 3$. The next period-doubling bifurcation occurs at $r = 3.4494$ when the two-cycle becomes a four-cycle. At 3.5440 the four-cycle becomes an eight-cycle, etc. Around about $r = 3.5699$ this period doubling ends and chaotic behaviour starts and is followed by 'windows' of chaos for larger values of r.

What do we learn from all that? We learn how a very small change in a parameter (namely r) can effect a transition for the eventual state of a population from a predictable value to a completely unpredictable (or chaotic) value. This remarkable result is due to the non-linear term in the original equation. Since the procedure leading to chaos is determined and is without any probability ingredients, we can (see section 5.1) apply the term deterministic chaos to it.

Return for a moment to the question of linear as against non-linear phenomena introduced in connection with the logistic equation. Here is a simple example. If you load a spring by suspending more and more weights from it, it will be extended more and more—a linear phenomenon, roughly speaking. The last weight may cause the spring to break. This is certainly a non-linear phenomenon.

Here is another example: a sandpile may be formed by allowing a fine stream of sand to fall slowly through a funnel to form a heap of sand. Another few grains and the pile will collapse. Such phenomena, like snow avalanches, are highly non-linear and it is difficult to predict their occurrence. The study of chaos explains this phenomenon of unpredictability which we are pursuing throughout this book.

5.4   Abrupt changes ('phase transitions')

Charles Darwin has to be the hero for this chapter which is of course greatly influenced by the ideas of biological evolution. The 19th century brought us three great theories of science: Darwinian evolution, the laws of thermodynamics (Chapters 2 and 4) and Maxwell's electromagnetic theory (not discussed here). Darwinism was all things to all men. The optimists saw in it a guarantee of continuing progress through future centuries. The pessimists found their own lives reduced to meaningless consequences of blind accidents. The statistical mechanical (section 4.1) discussion of the second law of thermodynamics suggests that entropy and disorder in an isolated system *increase* with time. Darwin's evolution hypothesis suggests the possibility of 'diversification' occurring, or a kind of 'breeding' among living things, with a resulting *increase* in order. Such a discrepancy can be at least partially resolved by observing that entropy increase holds for closed systems, while biological evolution holds for open systems. If these are driven far from an equilibrium state, they develop the following interesting characteristics.

(i) Several steady state solutions can exist and transitions between them are possible.

(ii) The steady states are not always stable against fluctuations and external disturbances.

An obvious example of an equilibrium transition is the melting of ice or the freezing of water: a slight change of temperature and the system changes to one with very different properties. We have varied a parameter and suddenly the roads become dangerously slippery! There are many other types of transitions, for example a magnetized medium can become demagnetized by external influences. These more general phenomena are called *phase transitions*. A cable may snap, a fuse may blow: these are also phase transitions.

I now give another example of a non-equilibrium phase transition, couched in the terminology of population dynamics, in order to make it more understandable. Consider, then, a typical member, P, of a population in which there are people who want to cause trouble; I call them agitators A. Certain interactions occur between members of the normal population, represented by P, and the agitators, represented by A. These interactions, (1) to (3) below, are a bit like chemical reactions, which I give in brackets.

(1) A member of the population may be converted to become an agitator $(P + A \rightarrow 2A)$.
(2) Some agitators may fall out among themselves and stop agitating as a a result. They become normal members P of the population $(2A \rightarrow P + P)$.
(3) An agitator may get disenchanted with agitation $(A \rightarrow P)$.

The following questions arise. What is the rate of growth of the number of agitators? Will the whole population be converted to their cause? This can be discussed if we associate with each of the three 'reactions' a constant representing the frequency of its occurrence. We find that in order to end up with a stable population with some agitators, the rate constant R (say) for the conversion rate (1) must be sufficiently large. The eventual number of agitators in the population increases with R (figure 5.7). On the other hand, if the conversion rate, R, is below a certain value, then we end up with no agitators at all: they have all reverted to being normal members of the population. Physics also finds uses for this model [5.10, 5.11].

If you plot the fraction of agitators in the population as a function of R (figure 5.7), the phase transition is found to occur at the corner of the curve. What is interesting, and somewhat unexpected, is that there is again a sharp value (this time of the quantity R) at which the transition occurs. If you regard the population without agitators as unstructured or symmetrical then, after the transition, you have a more structured population. The uniformity is broken and in that sense there is a loss of symmetry.

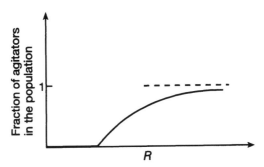

Figure 5.7 The steady state non-equilibrium fraction of agitators in the population, illustrating a simple non-equilibrium phase transition.

The 'reaction' (1) is of a type often found in chemistry. A chemical A encourages its own production. You start with one molecule of A and end up with two such molecules. This phenomenon is called *autocatalysis* and is of crucial importance for *self-organization*, for the generation of life, as well as for the production of chaos.

Here is a simple example. Imagine that we have an energetic electron in the conduction band of a semiconductor. Suppose also that it has enough energy so that in a collision with a valence band electron it can knock it into the conduction band. The result is a reaction of type (1): energy plus electron gives two electrons and a hole. This effect is called impact ionization and is important in semiconductor devices. Here it is revealed as an example of the ubiquitous autocatalysis.

5.5 Self-organization

Consider a physical system which is in equilibrium: nothing happens. At a later time the system is in a closely similar state. It is a rather boring situation which is a result of the processes during which equilibrium is attained. These processes have a direction in time and, except for the activity of demons (see section 4.2), they cannot normally be reversed for isolated systems. If the system is not isolated, but an open system (see section 5.3), there are many ways of inducing changes, for example by heating it up, by stirring it, by applying a magnetic field, etc. These processes are part of the subject of irreversible thermodynamics, which addresses the flows of energy and particles, as well as of other quantities, such as momentum, already met in section 2.2. Here the increase in entropy in the system or part-system of interest, must be expected to play an important role.

Here is an example. You can make measurements of heat loss per unit area from the human body as well as of skin temperature. The entropy generation rates can be calculated from this study of human body energetics. It is found to increase rapidly during the early developmental stages after egg fertilization. There then follows a decrease between the ages of 2 and 25 years, followed by a rather slow decrease during ageing. Thus a three-stage description, in terms of entropy increase, can be given of the human life span [5.12]. These changes seem to be due mainly to metabolic heat production.

Bifurcation phenomena indicate the possibility of bringing about organization by applying external forces to a system. Entropy is generated in such processes due to the supply of free energy from the outside and due to the flows generated by the applied forces. But the system of interest maintains itself, presumably at roughly constant entropy, so that it must do so by passing on entropy to its surroundings. There exists, therefore, a definite entropy production rate which the system lets fly into the outside world. Another way of putting it is to say that the system 'feeds on negative entropy'. This concept, made famous more than half a century ago in Schrödinger's discussion of life [5.13], does not do justice to the fact that living bodies must also give off heat. Recall the little verse on p 26, reminding us that a person is equivalent to two light bulbs burning continuously!

I shall now offer another example of a phase transition. Suppose that a material contains molecules which are themselves tiny magnets, but that these are randomly orientated so that the system is non-magnetic. It is called 'paramagnetic'. As it is cooled the interaction between the molecular magnets, which tends to line them up, becomes more important than the thermal agitation which keeps the magnets directed randomly. In due course, and normally quite suddenly, and thus via a phase transition, the magnets line up and the system becomes a magnet. It is magnetized in one of two possible directions: parallel or antiparallel to the applied field. Hence we have two branches on the left-hand side of the curve. In this way you have converted the non-magnetic material to a magnetic material merely by lowering the temperature. The simplest phase diagram for this case (figure 5.8) is not unlike figure 5.2, as it shows only one bifurcation. This type of behaviour is exhibited, for example, by iron and copper.

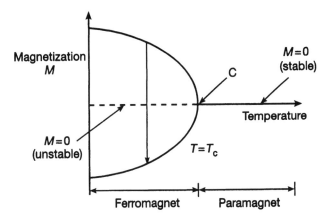

Figure 5.8 Magnetization as a function of temperature for a simple ferromagnet. C denotes the critical (or bifurcation) point.

It is remarkable that the application of an external influence is enough to bring about some organization, as has just been seen. This phenomenon of self-organization is a hint at a kind of synthesis between the random and the orderly. The two concepts are clearly not as far apart as might be expected. It reminds us of the term 'synergy', used nowadays by company chairmen (and others) to emphasize that their collaboration with another company is just the sort of activity which is of benefit to both. It is like 'collaboration' but with an extra accent on activity. In science it has given rise to the subject of *synergetics*, which aims to furnish guidelines for the study of cases when many subsystems act together to produce structure and functioning on a macroscopic scale.

The example from magnetism teaches us something else that is of wide interest. The solid can be perfectly uniform (if its lattice structure is neglected); there is no magnetization, no spatial direction is distinguished from any other. But once magnetization has been produced, the direction of magnetization gives a preferred direction: spatial symmetry has been broken. We can think of other such cases. For example, when water is cooled to form ice, the existence of ice crystals breaks spatial symmetry. Perhaps the most puzzling example is that in mechanics (classical or quantum) we have symmetry between the forward direction of time and its reverse. But in our macroscopic world the *forward* direction of time dominates, so that this symmetry is broken. We do not yet understand the reasons for this effect.

The concepts introduced in this chapter: self-organization, bifurcation, chaos, period-doubling, synergetics, non-equilibrium phase transitions, entropy production, autocatalysis, symmetry-breaking and so forth, have become popular during the last forty years or so, and must be credited at least in part to Ilya Prigogine (b1917; NL in Chemistry 1977) and his colleague Gregoire Nicolis (see, for example, [5.14–5.16]), Hermann Haken [5.17–5.18], Benoit Mandelbrot [5.19] and many other contemporary scientists. They are following the mathematical work of Henri Poincaré (1854–1912) and Alan Turing (1912–1954), to name only two of their distinguished predecessors. This work gives us a hint of how, very roughly, life might have originated. We have a series of bifurcations leading to self-organization, possibly initiated by an autocatalytic reaction. This development is non-linear and non-equilibrium.

5.6 Entropy is not always disorder

How can the fact of biological evolution and advancement be squared up with the second law of thermodynamics? The second law is often associated with the development of disorder, as we have seen (section 4.1). An important part of the answer, my first point, is that living systems are open. They are sitting in a larger environment. If we consider this larger system as closed for the purposes of our argument, then of course the entropy of this larger system has to increase; but it can still decrease in any small island contained within it. This allows for biological development within an overall entropy increase.

There is a second point. It is not actually true that entropy and disorder must always be linked, even though we have in earlier chapters gone along with this erroneous but widespread belief. I shall now break this link. This will further clarify the relation between biological development and thermodynamics. In order to discuss this matter sensibly, we need to be clearer about the concept of *order*, and for this purpose let us consider two ways of averaging four given ('original', *fine-grained*) probabilities (table 5.1). They can be probabilities for the states of *any* system you care to consider. I am interested here only in how you can manipulate these probabilities. The averages are ways of advancing from fine-graining to coarse-graining (section 4.4). I shall use averages of types I and II. It is crucially important that in type I the number of states of the system changes, while it remains the same in type II.

Table 5.1 Two ways of averaging four original probabilities. The table gives the probabilities.

	Extreme type I	Type I average	Original probabilities	Type II average	Extreme type II
			0.1	0.15	0.25
		0.3			
			0.2	0.15	0.25
	1.0				
			0.4	0.35	0.25
		0.7			
			0.3	0.35	0.25
Sum of probabilities	1.0	1.0	1.0	1.0	1.0
Actual entropy values	0	0.61	1.28	1.30	1.39

←——— Type I ———   ———— Type II ————→

Type I. We achieve coarse-graining by replacing two groups of two microstates each (column 3 in the table) by two macrostates, and assigning a probability to each which is the sum of the two probabilities in the group (column 2). I call this type I averaging. In the extreme case when the average is extended over *all* microstates (column 1), the resulting system has only one (macro)state, and the system has, therefore, to be in that state. This is tantamount to certainty within this coarse description. The entropy is therefore zero (see p 73). So this extreme case suggests that, even if you average only over groups of microstates, replacing each by a *single* macrostate, this process *lowers* the entropy. This is indeed confirmed by the actual numerical value (bottom line of the table). This type I averaging is occasionally used in statistical mechanics, and is very instructive. It shows, by going from left to right in the table, that if by some device the number of states is *increased*, then there is a tendency for the entropy to *increase*. Dimensionless entropies (S/k) are used in table 5.1 and figure 5.9.

Type II. Suppose we again have a system of four distinguishable microstates. A second way of coarse-graining is to lump pairs of states together and to associate with each pair two new (coarse) states, each with the probability averaged over the pair. Repeat this for the other pair of microstates, so that you have four 'macrostates' arising from the four microstates (column 4 in the table). The resulting new 'coarse' entropy, is greater than the original (fine) entropy. It is easy to see why: if you average over *all* microstates, you would have just one probability, the same for all four (coarse or macro) states (column 5 in the table). Because the probabilities add up to unity, this would give *equal* probabilities for all macrostates, i.e. the largest entropy. This extreme case suggests that coarse-graining of type II, even if it is not over *all* microstates, as in column 4, *increases* the entropy. This is again confirmed by the actual numerical value shown. Type II represents the usual form of averaging in statistical mechanics.

'The largest entropy'! To see how that comes about let us revisit our Himalayan community (p 71), where we saw that when everyone in the community is equally rich, then 'the happiness constraint' will stop more services being bought. The state of maximum happiness will have been reached. This corresponds to maximum entropy (column 5).

I shall now define *disorder*, *D* say (it is here a *technical* term), as the entropy of a system divided by the maximum entropy which it can have under the given circumstances. This number is zero if the entropy of the system is zero; it is unity if the entropy of the system has its maximum value. So *D* lies between 0 and unity. This gives me a chance to define *order*, Ω say, as $1 - D$. Then zero disorder means maximum order Ω (= 1), which is reasonable. Maximum disorder *D* (= 1) means order (Ω) is zero, which is also reasonable. Thus Ω also lies between one and zero.

A simple illustration will make us more familiar with these quantities. Consider a system with two equally probable states. Its entropy is then the maximum compatible with two states (because the states are equally probable). In our Himalayan village analogy it corresponds to a population of just two equally rich people. Happiness can no longer increase, just as entropy can no longer be increased when the states are equally probable. Suppose we now find an improved apparatus, and that with its aid manage to resolve the upper state into *r* states, equally spaced about the upper energy level, so that the total energy is unchanged. The probability that the system is in its lower energy level is assumed to remain at one half. A little thought shows that this process corresponds to the *reverse* of type I averaging. Hence the entropy goes up. Calculation shows that the maximum entropy goes up as well, but more rapidly, yielding a drop in disorder. The result is that both order and entropy go up together as *r* is increased (figure 5.9, see [5.20]). This is important for the analysis of

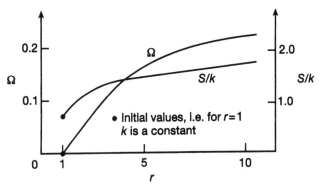

Figure 5.9 Entropy and order of a two-level system as a function of the number of levels (*r*) into which the upper level is resolved.

life processes since biological systems exhibit an increasing number of distinct states during growth. Thus order, as defined here, can go up as the entropy goes up.

Biological applications of these ideas are possible (see, for example, [5.21–5.23]).

5.7 The origin of life

I now come to the nature of living systems from the viewpoint of a physical scientist, using concepts introduced in this chapter. They are open systems, based on carbon (*carbon life*), which can maintain themselves in a state far from equilibrium; and they can grow and multiply, using a flow of energy and matter supplied by the environment. Silicon is chemically similar to carbon, so that *silicon life* may one day emerge from science fiction into science.

Here are some requirements for living systems [5.24]. A living system must be able to:

(i) Manufacture its own constituents from the materials available in its surroundings.
(ii) Extract energy from the environment and convert it for its own use.
(iii) Insulate itself to keep control of its exchanges with the outside.
(iv) Regulate its activities to preserve its organization in the face of environmental disturbances and fluctuations.
(v) Help to accelerate the many chemical reactions needed to support its activities.
(vi) Multiply so as to propagate itself.
(vii) Arrange its biological processes so as to guarantee reproduction.

This topic has attracted much attention and biologists, physicists and philosophers have discussed it for many years. Consider requirement (vi) as an example.

Suppose a living organism is in some (quantum mechanical) state in contact with a nutrient environment, the whole being enclosed in a box. I ask: does there exist a reasonable (quantum mechanical) interaction between the two which would lead *with certainty* (*assumption A*) to a new state of the system, which contains two similar living organisms? This is the problem of self-reproduction. The relevant interactions are not known exactly—certainly not in this very complex case. This has not deterred people from discussing it. They have produced successful theories involving statistical averages over a reasonable *class* of interactions (*assumption B*). This was done in a particular case, and the shocking result was found that self-replication appears to be highly unlikely [5.25].

If taken seriously, such a conclusion would wreck quantum theory! After all, life exists! Fortunately a weakening of the above assumptions A and B is possible, and by this means we can recover the possibility of life, on the basis of quantum mechanics [5.26]. We have to be satisfied with a mere probability of life being produced (assumption A). Further, we have to make the class of interactions considered somewhat wider (assumption B). We can then find the necessary result: a possibility of life. This is fortunate not just for life, but for quantum mechanics!

The question of the probability of life emerging from some initial 'soup' is still of interest nowadays, certainly in connection with the probability of life on other planets and elsewhere in the universe, and so this question continues to attract attention [5.27].

A completely different attack on the problem of the origin of life is by means of laboratory experiment. Take something like the atmosphere as we believe it was constituted about four thousand million years ago when life began to emerge. (The earth itself is believed to be about four and a half thousand million years old.) Run an electric discharge through it, simulating lightning, and see what you find. Amino acids were found—a possible first stepping stone to life via the production of proteins. This way of understanding the origin of life [5.28] was taken further by Stanley Miller in Chicago, and it was thought that nucleic acids and enzymes can also be made in this manner. However, this view is no longer accepted.

Another idea is that viruses and bacteria developed in outer space and were brought here by comets (Fred Hoyle and Chandra Wickramasinghe). Further, microbial spores may have polluted the interstellar medium from life in other regions since under favourable conditions they can have a survival rate of the order of millions of years [5.29]. Francis Crick suggests that life arrived here in a space-ship sent by beings who are much more intelligent than ourselves [5.30]. These phenomena are called panspermia, originally proposed by the physical chemist Svante Arrhenius, and are taken quite seriously in some quarters. But this ignores the scientific problem of how life originated from dead matter.

I have been thinking in reductionist terms. We try to explain biological effects in terms of physical principles. This is a good working hypothesis, but the fact that it is a hypothesis shows us again the presence of a real incompleteness in our understanding.

The mechanisms discussed above have all been investigated and it is difficult to pin our hope for at least a preliminary understanding of life processes to any one of them. Yet autocatalysis, met in reaction (1) in section 5.4, stands out. Chemists have actually made a synthetic molecule which can make copies of itself. One compound (amino adenosine triacid ester, but the name does not matter here) pulls in molecular fragments to make a copy of itself. It is different from normal biological replication in not requiring the assistance of an enzyme. There is hope for progress in this direction [5.31–5.33].

When things are too difficult for us, we can always emphasize the point by the invention of a demon! Following a recent suggestion [5.34], I shall appeal to the 'Eigen demon' D7 (see p 78) which can convert inanimate to animate matter. We humans cannot achieve this yet.

If we think back to anti-diffusion and anti-heat conduction, these processes change completely when we reverse the time axis. We say that they are not time-reversal invariant, or not T-invariant. But they occur frequently: the most elementary processes which bring them about are particle collisions, and these are *not* violated by reversing the direction of time, as explained in section 4.2. They *are* T-invariant. Let us encapsulate this idea in a new term and say that diffusion and heat conduction are *weakly T-invariant*. The definition is: 'a complex process is weakly T-invariant if its time inverse, though perhaps improbable, does not violate the laws of the most elementary processes in terms of which it is understood'. Of course *all* macroscopic processes in *inanimate* matter are weakly T-invariant. But we have not yet established scientifically that the laws of physics and chemistry are adequate for the explanation of living matter, of consciousness, of purpose, etc. Thus we arrive at a possible *definition of life* as a macroscopic process which violates weak T-invariance [5.35]. This definition focuses attention on the reduction problem of biology, namely the question of whether living systems can be described fully in terms of physics and chemistry, i.e. in terms of weakly T-invariant processes. Thus we have here a definitions of life which can at least make us think!

5.8   Summary

Our story contains the illustrious names of Boltzmann, Carnot, Clausius and Kelvin, who showed in the second half of the 19th century that the entropy of a finite isolated system tends to increase. Broadly this means that a physical system, if left to itself, tends to get more disordered: ice-cream with hot chocolate sauce tends to become a lukewarm mess. Why 'tends to' and not an unequivocal 'does'? The reason is that for small systems the entropy is liable to oscillate, and for microscopic systems it is a parameter which is normally of little interest. There is the possibility that the ice-cream will be cooled further, and that the chocolate sauce may boil in virtue of some conspiratorial re-arrangement of energy (*not* of a creation of energy) among the atoms. We know from experience that this is highly unlikely.

By going to the limit of an infinitely large system these possibilities disappear: entropy increases uniformly for an isolated *infinite* system. Of course, you can object that the universe may not be big enough to contain an infinite system; so you have these infinite systems merely as a helpful calculational device. Even this is not good enough, though, since the gravitational forces will cause an infinite system to collapse, and so equilibrium is not attained. Anyhow, keeping to finite systems, the direction of time enters physics by way of this tendency for entropy increase. This situation was finally summed up as 'entropy—the arrow of time' in Eddington's Gifford lectures [1.1]. In the following decades quantum mechanics scored success after success, explaining atomic collisions in gases, the properties of solids, stellar interiors, etc. It was assumed in Max Born's Waynflete lectures [5.36] that a deterministic development in time, given by classical mechanics, had been replaced by quantum mechanical ideas that microscopic physics could not go beyond statistical statements.

Determinism was at least temporarily dead in microscopic physics, and consequently very sick indeed in the rest of physics, chemistry and biology.

Although not mentioned by Born, Schrödinger had in the meantime used both thermodynamics and quantum mechanics in his seminal 'What is Life?'. In fact, a guilt complex was developing among us physicists that we had not squared up the ordering processes of biology—the emergence of new and more highly developed species—with the disordering tendency of entropy to increase. The centre of interest was in fact shifting to the effect of the environment on systems: the open system moved to the centre of the stage. Schrödinger had already emphasized that living systems keep their entropy down by losing some of it to their surroundings. This general feeling led to interesting discussions as to the way in which quantum mechanics could handle this problem (see above).

This phase of looking for biological insights by using the general principles of physics, without, however, letting the mathematical analysis burst out in new directions, was concluded with four volumes published under the editorship of C H Waddington in 1968–1972 with the title '*Towards a Theoretical Biology*' and in a book entitled '*Theoretical Physics and Biology*', edited by M Marois (1969).

In the meantime evidence had accumulated for biological oscillations. Nicholson (1954) had found a periodic time of 30 to 40 days for the numbers of a blowfly *Lucilia* in an experimental population to which an unlimited food supply was available; the moth *Bupalus* exhibited population fluctuations with a period of about six years in pine forests in Germany (1949); Goodwin (1963) had discussed sustained oscillations in controlled biochemical systems incorporating feedback; in the USSR Zhabotinsky and co-workers had studied oscillatory processes in chemical and biological systems. In another direction Fröhlich (1968) had suggested that the energy supplied to biological systems might in part be stored in an orderly fashion in some kind of *Bose–Einstein condensate* (see section 6.10).

By the 1970s interest had shifted to systems in a steady state maintained by absorption and rejection of appropriate streams of energy. Such systems often have their steady states far from their equilibrium states, are subject to non-linear equations, and can be taken from one regime to a completely different regime by the manipulation of external conditions, such as the electric field in physical problems or food supply in population studies. These induced sudden changes were referred to as non-equilibrium phase transitions. A normal phase transition, such as condensation on a car wind screen, is also sudden, but is a near-equilibrium change.

Attention was now focussed on the non-equilibrium cases which make life so happy: the sail which suddenly flutters in the wind, the waves which break on the beach, etc. Inspiration for the understanding of some of these phenomena came from population dynamics, considered by the American Alfred Lotka many years earlier [5.37]. After all, you could easily find oscillations in populations: too many predators lead to a shortage of prey and hence of food. So the predator population decreases. This enable the prey to establish themselves more strongly, and the whole cycle is repeated. Of course, if the interaction between predator and prey is of a special kind, a completely different regime can occur: the prey can die out, followed inevitably by the dying out of the predator for lack of food.

The richness of structure in these non-equilibrium systems brought about by non-linearities in the equations was very impressive, and many reviews and books were written in the 1970s, in the stimulating border territory between physics and biology. The problem of evolution was also covered: *Le Hasard et la Necessite* (Monod 1970), *Complexity in Ecosystems* (May 1973), *Stabilite Structurelle et Morphogenese* (Thom 1972). Selection was illustrated by the results of randomness in board games (*Das Spiel*, Eigen and Winkler 1975). Self-organization was now discussed more widely by combining cycles of chemical reactions into 'hypercycles' (Eigen and Schuster 1978) and in 'Self-organization in non-equilibrium systems' (Nicolis and Prigogine 1977). These varied topics are related to each other by the mathematics used—often referred to as catastrophe theory.

In this impressive list the word 'self-organization' enters with special reference to life. In a narrow sense it refers to the capability of certain forms of matter to give rise to self-reproducing structures under fixed external conditions and internal interactions. An interesting example is the autocatalytic reaction in which a steady non-equilibrium state is maintained by a chemical which encourages (strictly, catalyses) its own production. Could it be that some new laws of physics are waiting to be discovered and would enable us to make more sense of life processes in terms of physics?

We should retain one key idea from the work of the past decades. The conceptual stranglehold on the evolution of the universe due to the second law of thermodynamics has been broken. Increase of disorder, the end of life, the heat death—they all apply to closed systems which approach equilibrium. But is the universe approaching equilibrium? Or will *gravitation* (neglected in standard thermodynamics) prevent it? Also, the systems *within* the universe interact and are not closed, thus opening the way to novel and unexpected non-equilibrium states as evolution proceeds. Speaking rather vaguely, Darwinian optimism is winning the upper hand over the pessimism of the physicists of the end of the 19th century!

After this exploration of statistical effects and the borderland between macroscopic and microscopic effects, we turn next to the physics of the very small, continuing the story started in section 3.9.

Chapter 6

Now you see it, now you don't Quantum theory: science and the invention of concepts

6.1   Introduction

We now come to the most successful physical theory of all time. Fifty years have elapsed since the 'new' quantum theory emerged from the 'old' quantum theory of Niels Bohr. It has been used to explain the properties of semiconductors and transistors and has given rise to the field of quantum electronics; it has been applied to gases, to stars and to atomic nuclei, and it has never failed; put more carefully, the inaccuracies of the results have never exceeded values regarded as reasonable.

I have already covered important aspects of quantum theory in earlier chapters: I have talked about energy levels of electrons in atoms in section 3.8, and at the end of that section we noted that a transition from the 'old' Bohr quantum theory to the 'new' quantum mechanics took place around 1926. We met antimatter in section 3.9 and the exclusion principle in section 3.8. These are newly invented concepts and belong to quantum mechanics.

I have explained a little bit of what quantum theory *is*; next let us see how it *arises*.

6.2 Quantum mechanics: the elimination of unobservables

In the process of finding laws and creating order it often happens that scientists introduce a situation, or a concept, which they think exists, or is feasible, but which has not actually been shown to exist experimentally. Such concepts or situations are called 'unobservables'.

For example, in the 19th century there was a dispute: the notion of an atom was regarded by some (see section 3.5) as an unobservable which should be eliminated from physics considerations, while others believed in its usefulness. Sometimes unobservables indicate that theoretical thinking is ahead of experimental work. Most important are those unobservables which enter physical theory unheralded and unsung, simply as part of the physicist's normal thinking equipment. The elimination of such unobservables can lead to profound advances (see point (ii) below).

Although we can *now* approach quantum theory from this point of view, by hindsight as it were, historically it was forced on us by 65 years of experiments and theory. Still, it is very useful to pursue this hindsight type of approach for a moment, for it can surely be argued [6.1] that in many cases a strictly historical treatment can make it harder to understand what is going on.

I offer you now two unobservables which ingratiated themselves into the physics of the 19th century, strictly speaking without justification, and are queried by some contemporary physicists.

(i) Consider the motion of a ball. At every point on its path it goes through a position in space with a definite speed. So it is not unreasonable to suppose that this also holds for atoms and particles as well as for trains and aeroplanes. It must be right—you can *see* it with your own eyes. But this is only common sense, and that is known to be unreliable. Could it not be that, as far as an elementary particle is concerned, it passes along by being destroyed and recreated an innumerable number of times, and so does not execute a continuous curve at all? This is just an idea to shake our confidence in common sense. However, following Pythagoras and Henri Poincaré, it could be 'that space and time are granular, not continuous' [6.2]. This is a possibility, but it is not generally accepted.

(ii) To find the velocity of a particle, put a stop watch at two places on its path, and divide the distance between them by the time taken. The result is only an *average* velocity for this time period, and it is impossible to associate *one* position or *one* instant with this average. Thus: if the (average) velocity is known accurately, then both position and time are uncertain. The converse also seems sensible: if the position is known accurately, what can the velocity possibly be? This kind of talk goes back to the Greek philosophers of about 500 BC. Zeno the Eleatic claimed then that an arrow at any point on its path is effectively at rest and there is then nothing in the system presented to us at that instant to tell us that the arrow is in flight. While Zeno's argument may be correct, his assumption—that a particle can be at complete rest—violates quantum theory (p 154). Though nobody now accepts his inference that motion is impossible, we see that his argument does suggest some kind of competition between position and velocity (or position and momentum). Such pairs of observables are called *incompatible* and lead to the famous *uncertainty relation* of quantum theory.

Within the framework of quantum theory, in measuring the properties of a particle we find the minimum experimental error in the position measurement multiplied by the minimum experimental error in the momentum measurement must always exceed a certain numerical value, $h/4\pi$, where h again denotes Planck's constant (see p 46). This is the Heisenberg uncertainty principle which applies to measurements in each of the three directions of space. It holds, furthermore, for many other pairs of physical quantities, e.g. energy and time. The uncertainty stipulated by the principle is negligible in normal life, but it yields key constraints in the quantum domain. Our knowledge of physical variables such as time, energy, position in space, etc, which were defined centuries ago within the framework of classical physics, cannot be known as completely as had been thought. This is intrinsic incompleteness brought about by human idealization.

If judged by Newtonian, i.e. classical, physics, the uncertainty principle suggests, therefore, an incompleteness in quantum mechanics: we cannot know things we believed classically we ought to be able to know. This is a view which was favoured notably by Albert Einstein and by David Bohm. You could call it the C view. But suppose we immerse ourselves completely in quantum theory, and regard the current views as correct for the time being. Let us call this the Q view. Then the apparent shortcomings which we have observed are not an incompleteness, rather this is just how things *are*: the incompleteness is apparent rather than real. Thus there emerge two different interpretations of quantum theory: the C view, which suspects an incompleteness, and the Q view which does not.

The uncertainty relation implies another notion, namely that knowledge of one aspect of a physical system precludes you from having as detailed a knowledge as you might have expected of another aspect of the system. This is the *complementarity principle* of Niels Bohr, and it applies to pairs of incompatible observables. In later years he applied it more generally to pairs like life and matter or love and justice, giving it a more philosophical flavour. The principle is essential to the so-called Copenhagen interpretation (which is where Bohr worked) of quantum theory. It tells us that a system may be described by using one or another set of observables, and the results, while of course not identical, should be regarded as complementary. The apparatus used for the measurements, and the system measured, have here to be treated separately in any theoretical discussion. Furthermore, the actual values of the variables involved are regarded as indeterminate until a measurement is made.

The result of the measurement is that the mathematical object which represents the *state* of the system, namely the so-called *wavefunction*, changes upon measurement to a state which reflects the result of that measurement. The wavefunction is often said to *collapse* to that state. Thus we cannot say much about quantum mechanical magnitudes until they are measured. As John Wheeler, the well-known American physicist, likes to put it: 'A phenomenon is not a phenomenon until it is a measured phenomenon'. Note all the concepts specially invented for quantum theory.

Werner Heisenberg (1901–1976; NL 1932) derived the uncertainty relation by using an imagined microscope in order to locate an electron. Since light (i.e. a photon) has to bounce off the electron and into the microscope, we can say upon noting the photon 'Ah, here is an electron'. In reality the procedure is more difficult, but then Heisenberg was considering only a 'thought experiment'. The electron can be located with fair accuracy if the distance separating successive crests of the light wave lie close together. This distance is the *wavelength* of the light, so that a short wavelength is required. Indeed, the smaller the object the shorter the wavelength has to be to detect it. Short wavelengths always means high frequencies, and this means high energy (section 2.3). So to find a small object you have to hit it with photons of high energy; the resulting recoil makes the momentum of the object uncertain. In this way we gain a qualitative grasp of the uncertainty principle involving position and momentum (see e.g. [6.3]). Analogous considerations apply to other observables. Thus different spatial components of the spin of a particle (p 45) are also incompatible.

If we write '*e*' for energy, then this symbol stands normally for all possible values from zero upwards, or perhaps for all values between certain limits. But only a discrete set of values is possible for electronic energies in atoms (p 45). What do we do about that?

We need a new mathematical object E so that a classical–quantum *correspondence* (p 46) can be expressed by an arrow:

a normal variable $e \rightarrow$ a new mathematical object E.

The symbol E must have hidden in it a sequence of numbers representing, according to some mathematical scheme which need not concern us here, the allowed electronic energy levels.

Similarly, if the classical position and momentum of a particle are denoted by x and p, their quantum counterparts may be denoted by X and P in an analogous manner. The rules for manipulating these new mathematical objects are not needed here, but they imply that XP and PX need not be the same. The reason? X and P are now not just numbers like 3 and 5 for which 3×5 is the same as 5×3. They are these new mathematical objects. If XP and PX are not the same, the objects X and P are said to be *non-commmuting* and the variables to which X and P correspond are said to be *incompatible*. The result of the measurement is then that the object which represents the *state* of the system, called its *wavefunction*, 'collapses' to the appropriate state corresponding to the result of the measurement.

The new symbols E, X, P, etc, are sometimes called *matrices* and sometimes *operators*. The numbers required to specify the energy levels turn up as so-called *eigenvalues*. These mathematical details are not for us here, so let us pass to another approach to the quantum theory.

6.3   Wave mechanics: the optics–mechanics analogy

I have talked about mechanics using the idea of particles. However, since the time of Pierre Fermat (1601–1665), famous for the recent proof of his *last theorem*, an analogy between mechanics and optics has fascinated people (figure 6.1).

For example, the simple reflection of a light ray from a mirror follows the same laws as do particle collisions on a billiard table: the angle of incidence (i) equals the angle of reflection (r) (figure 6.2). More sophisticated analogies were also known. For example the *actual optical path* of a light ray connecting two points was shown to always be the shortest of the *possible* paths. This is Fermat's principle of *least time* and is part of what is called *geometrical* or *ray optics*. (In a uniform medium the optical path is just the ordinary path, but if the ray passes through a variable medium, a correction has to be applied for each increment of the actual path to turn it into this optical path.)

Although Maupertuis did not state it very clearly, nonetheless his principle of least action (figure 6.1) was found to be a close analogue in mechanics to Fermat's principle in optics. Why should *minimum principles* ('least' time, 'least' action) be so important? There have been many speculations on this subject, from Maupertuis, who envisaged a theological connection, to the advent of modern field theories based on Lagrangians, which attained an early importance by virtue of such principles. I leave the reader to contemplate this essentially unsolved question.

In the 19th century the 'wave' aspects of optics, had also been studied. When light passes through holes or passes obstacles whose size is similar to that of the wavelengths of the light used, geometrical optics has to be replaced by a subject which takes account of the wave nature of light. This is called physical optics. The colours we see in soap bubbles and on thin films of oil involve thicknesses which are of the order of wavelengths. Such phenomena can be analysed only with the aid of physical optics, invoking optical interference, explained below.

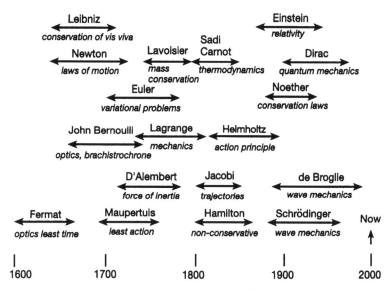

Figure 6.1 Well-known scientists and their times.

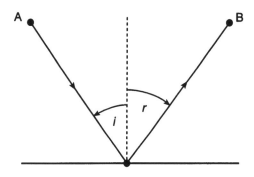

Figure 6.2 Equality of the angles of incidence (*i*) and reflection (*r*).

Let me now consider the wave concept of light, which we have already met (section 2.3) when I talked about the colours of the rainbow, and later when I talked about the colour force (section 3.9). Newton (1642–1727) favoured a particle picture for the propagation of light, at least partly for the reason, noted above, that light and particles obey the same laws of reflection at surfaces. So he argued from his knowledge of the behaviour of particles, that light, too, consisted of particles. Due to the gravitational attraction between these particles and the medium through which they move, he also inferred correctly, though in disagreement with the facts, that the denser the medium, the faster light must move in it.

At the same time Christian Huygens (1629–1695) suggested that light consisted of waves. You see, strictly speaking, light does not produce a sharp shadow, as you might think at first sight, and as you would expect from Newton's particle theory. If you look carefully, the edges of a shadow are a little blurred. Further, if you shine light through a small hole you finds that it bends a little around corners. This is the *diffraction* phenomenon, easily demonstrated for sound (figure 6.3). It is least important if the opening is bigger than the wavelength of the sound, and becomes more so as the opening is decreased in size. Diffraction suggests that light consists of waves, and as such requires a medium to move through, just as sound waves need air, and water waves need water, to exhibit their wave motion. It was more difficult with light, as it reaches us from the stars and the sun by traversing vacuum. To complete the picture, therefore, an omnipresent *æther* was postulated, which extended throughout space. On this basis you

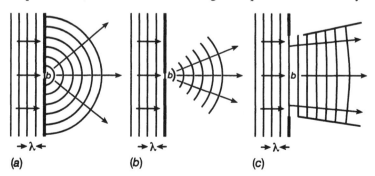

Figure 6.3 Diffraction: sound waves diffracted by apertures of different sizes.

could show that the denser the medium the slower the waves should move, in contrast to what the particle picture had incorrectly suggested (see above paragraph). We conclude that wave optics was able to explain phenomena such as interference and diffraction, which geometrical optics just could not handle.

More difficult is the question: do scientists believe in the existence of an æther? A popular answer is 'no': optical experiments at the end of the 19th century showed conclusively that motion through this æther just could not be detected, so that it was simplest to suppose that it does not exist. But nowadays particles are regarded as 'excitations of an appropriate quantum field'; for example photons are excitations of the electromagnetic field, electrons are excitations of the electron field, etc. At this level the æther can be regarded as being as rich in objects which it contains as the vacuum (see p 152).

Table 6.1 formalizes the optics–mechanics analogy. What corresponds to physical optics in mechanics? Clearly a subject in which particles have wave-like properties. This is provided by the wave formulation of quantum mechanics, called *wave mechanics*. People who already know the various aspects of optics could call this new mechanics, 'the physical optics of classical mechanics', or they could call the old mechanics [6.4], 'the geometrical optics of wave mechanics'!

Table 6.1 Analogies between optics and mechanics.

	Optics	Mechanics
Rays or particles	Rays in a series of straight lines. Fermat's principle. Geometrical optics.	Motion in a series of straight lines. Least action principle.
Waves	Interference, diffraction, etc. Physical optics.	?

As to interference, figure 6.4 shows two waves A and B which are propagating at the same time. The two waves add to yield wave C. At the coordinate marked with an arrow wave C has reached a larger amplitude, and at the coordinate marked with a cross they have cancelled out. We say that the waves have interfered constructively and destructively respectively. Such interference is possible over long distances if the waves are coherent (p 146), i.e. they are simply related as in figure 6.5. They then have the same wavelength. Let us imagine an arrangement in which light from one vertical rectangular slit hits two similar slits so that two coherent waves can emerge. Although the waves are cylindrical, they are shown as circles of maximum and minimum displacement. When the maxima from the two slits overlap we have reinforcement of the waves and maximum illumination. When a maximum and a minimum displacement come together we have destructive interference and no illumination (figure 6.6). On a vertical screen you then see bright and dark lines alternately. This is a characteristic interference pattern which cannot be explained by geometrical optics.

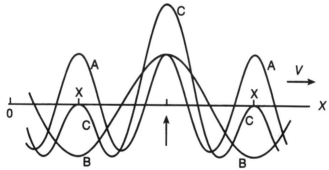

Figure 6.4 Interference: waves A and B acting together yield wave C. (Reproduced from figure 38.1 in [6.5].)

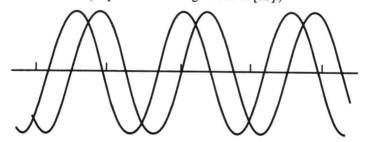

Figure 6.5 Coherent waves. They have the same wavelength. (Reproduced from figure 38.2 in [6.5].)

Note that this result can be generated by starting with a small number of photons produced typically in a low-power *laser*. They hit the screen, making little dots of light. As the intensity or the exposure time is increased, the interference pattern emerges. Thus the arrival of the individual photons must be governed by quite definite probabilities.

Somewhat similar conditions hold for sound waves. The two prongs of a tuning fork when sounded act as identical sources of sound. If the fork is placed close to the ear when sounded, and slowly rotated about its handle, then four positions can be found for which no sound is heard. This illustrates interference of the waves emanating from the two prongs of the fork.

Again, when two stones are dropped into water at the same time, two sets of concentric wave patterns are found and an instantaneous photograph shows the lines of destructive interference (figure 6.7).

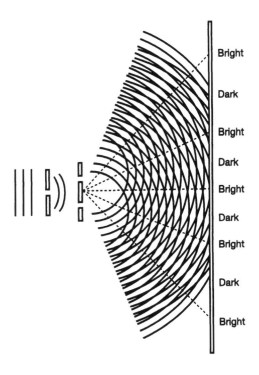

Figure 6.6 The incoming wave, three apertures (which produce two identical sources) and a screen with alternating bright and dark regions. (Reproduced from figure 33.3 in [6.5].)

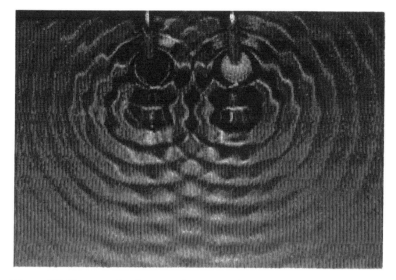

Figure 6.7 Photograph of pattern due to two similar sources of water waves, showing the destructive interference regions.

6.4 A brief history of the new mechanics

Photons not only have the obvious wave aspects but also particle aspects (see above). Could it not, therefore, be that, similarly, particles such as electrons can have wave aspects? That something fairly drastic was needed can be seen from the fact that nobody really understood why the Bohr orbits of electron in atoms were stable. A negative charge rotating around a positive one is expected in due course to fall into it, radiating away energy (see also p 43). Where can you look for help? The occurrence of integers in the form of Bohr quantum numbers suggested the relevance of waves.

Why? The 'music of the spheres' was a concept known to Johannes Kepler (1571–1630), who wondered about the notes produced by the planets as they moved in space. This idea may have come from the observation that a note was produced when a stone was whirled about a stick to which it was attached. Also, the Pythagoreans had noticed that harmonious sounds are produced by two strings in equal tension if their lengths are in the ratio of simple integers. It was also known that a violinist can touch a string at the centre so that this point becomes a 'node' (a point of zero displacement). The resulting frequency (figure 6.8) is twice that of the 'fundamental', which occurs if there are nodes only at the ends of the string (figure 6.8(a)). If the string is touched at one-third of the distance from the end (figure 6.8(c)), the frequency of vibration is three times that of the fundamental. These characteristic frequencies are called *eigenfrequencies* and are in this case simple multiples of the fundamental. They are closely related to the eigenvalues noted on p 127.

Lastly, consider a 'singing' wine glass. The vibrations are induced by rubbing the rim gently. The wave has to come back to link up with itself (figure 6.9(a)) to produce a note which does not change with time. If it does not link up (figure 6.9(b)) there can be no note! This note is an *eigen*-frequency of the glass.With more energy input one can of course excite the next possible higher note. As it has twice the frequency, it is the octave. Again you arrive at a characteristic frequency.

I have said enough to give weight to de Broglie's remark in his Nobel lecture [6.6]:

> '...the stable motions of the electrons in the atom involve whole numbers, and so far the only phenomena in which whole numbers were involved in physics were those of interference and eigen vibrations. That suggested to me that electrons themselves could not be represented as simple corpuscles either, but that a periodicity had to be assigned to them too.'

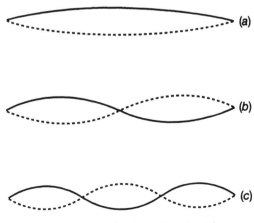

Figure 6.8 Vibrational modes of a string.

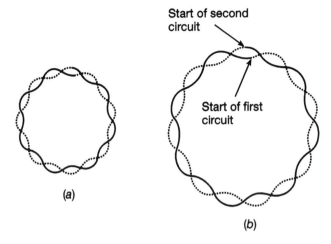

Figure 6.9 Vibrational mode for the top of a wine glass.

The revolutionary new idea from Louis de Broglie of Paris (1892–1987; NL 1929), which I have tried to make plausible in the above paragraphs, was to associate waves not only with photons, but also with particles of matter. Experiments were already to hand in 1923 of maxima and minima in the electron beam intensities reflected from metallic surfaces, but their interpretation as diffraction effects due to electron waves was still missing. This was achieved in 1927 by C J Davisson (1881–1958; NL 1937) and L H Germer (1896–1971) and, in independent experiments by G P Thomson (1892–1975; NL 1937). He, the son of J J Thomson who showed that the electron is a particle, thus helped to establish that it was also a wave. Davisson and Germer were helped by an accidental 'heat treatment' of the nickel surface which they were bombarding with electrons. This heat treatment was due to the explosion of a flask of liquid air in their laboratory! The result of their findings was described by Davisson in his Nobel prize lecture [6.6]:

> '...Not only had light, the perfect child of physics, been changed into a gnome with two heads— there was trouble also with electrons. In the open they behaved with admirable decorum, observing without protest all the rules of etiquette..., but in the privacy of the atom they indulged in strange and unnatural practices....'

such as being wedded to Bohr orbits!

The wave nature of electrons, and hence of other material particles, having now been accepted, the more formal development of a new mechanics had to follow. In line with the ideas of p 126, it arrived in 1925 as *matrix mechanics* in the hands of Heisenberg (1901–1976; NL 1932) on the one hand, and of Max Born (1882–1970; NL 1954) and Pascual Jordan (1902–1980) on the other. Following the ideas of p 130, it arrived, in a second form, as wave mechanics in 1926, in the hands of Erwin Schrödinger (1887–1961; NL 1933).

Only a little time elapsed before Schrödinger established the equivalence of the two approaches, one algebraic, the other using differential equations. Other theories existed, notably that of Paul Dirac (1902–1984; NL 1933), but the main thrust of later work resided in the literally thousands of application of the now unified quantum mechanics. Thus Japan has hosted several symposia (the fifth in 1995 [6.7]) on the foundations of quantum mechanics 'in the light of new technology'. They have dealt with topics which are so advanced that even the titles of the papers are hard for an outsider to interpret. One of the easier ones deals with 'laser (see Glossary) cooling'. Since radiation exerts a pressure rather like a gas, the cooling can be achieved by utilizing the pressure of laser radiation in such a way that the velocity of an atom in the system of interest always 'feels' a force opposed to its velocity, thus slowing it down. Slower movements imply lower temperatures. Other topics include *quantum gravity*, an attempt to combine relativity and quantum theory. This is needed because relativity theory, which deals with gravitation, uses events and physical objects at *points* in space and time which quantum theory cannot accommodate. Its physical objects cannot be reduced beyond small volumes or lines. That was a reason for introducing *strings* to represent particles as possible components in quantum gravity.

6.5   Wavefunctions and probabilities

Quantum theory characterizes a quantum system (i.e. one in which quantum effects are dominant) by a so-called wavefunction. This is a mathematical function which depends on the time and on the position in space. It also defines how a fixed and more or less fully specified system changes with time. The wavefunction is obtained by the solution of the so-called Schrödinger wave equation. That is, of course, different for different systems. This is so because you have to insert into the Schrödinger equation knowledge of the particular energetic properties of the system. The wavefunction is the basic key. So, suppose we have that function for a certain system. How does that help? It enables you to calculate the probabilities of finding the system in any of its known states of interest by using certain standard rules. That is the end product of the calculation. The states of interest can refer to states of specified momentum, angular momentum, etc.

Quantum states can be independent of time. These are stationary states. Typically they arise (figure 6.9) if the end of the first circuit coincides with its beginning, and therefore with the beginning of the second circuit. Then there are time-dependent states, but we need not discuss these here.

A wavefunction can normally be decomposed (in a sense that also need not concern us here) into components. The state it represents can therefore be regarded as a *superposition* of other states. If these are states of different positions, for example, then the original state participates in both of them. In that sense the particle is then in two places at once. We already knew about superpositions from acoustics, since most instruments, even harps, violins, horns, etc, do not sound a pure note but several pure notes at the same time, giving each instrument its individuality. This is somewhat analogous to a superposition of states. (A technical point: the possibility of superposition is a result of the so-called *linearity* of the wave equations, which holds for sound as well as for wave mechanics.)

Here is the outline of a simple example to fix ideas. Suppose we are interested in the emission of an electron from some system like a metal which can be represented in a simplified manner by a box. We specify the initial state, the final state of the electron being outside the box. The energetic properties of the box are inserted into the Schrödinger equation. This is solved for the wavefunction, from which we can then calculate the emission probability. That is the key result. Some additional routine matters have then to be attended to:

finding how the resulting probability depends on the shape and size of the box, on the initial state of the electron, etc, and then a graph can be plotted of the probability as a function of these parameters.

But *how* did the particle get out of the box? Since particles can be represented by waves, they can get past obstacles like walls of boxes, as we have already seen in the case of diffraction. This quantum mechanical *tunnel effect* works best for small particles (like electrons) and thin walls. For heavy particles the wavefunction outside the box is negligible, and classical physics reigns. For particles which are small enough, the wavefunction outside the box normally becomes stronger as one increases the energy of the particle, and so the particle has a better chance of getting out of the box.

Incidentally, the tunnel effect also plays an important part in the famous fusion reaction which, it is claimed, will one day produce cheap electricity. In this reaction we need to get two deuterons (nuclei of deuterium) to interact to produce (i) a triton (nucleus of tritium, see table 6.2), (ii) a proton and (iii) lots of energy. To get the deuterons close enough to each other so that they can react, the temperature has to be very high (see table 2.1). They can then get close by virtue of the tunnel effect.

Table 6.2 The simplest elements.

Element	Isotopes
1H, hydrogen (one proton plus one electron)	2H, i.e. deuterium (1 proton,1 neutron, 1 electron) 3H, i.e. tritium (1 proton, 2 neutrons, 1 electron)
2He, Helium (two protons plus two electrons)	3He, i.e. 2 protons, 1 neutron, 2 electrons 4He, i.e. 2 protons, 2 neutrons, 2 electrons

The notation which puts the atomic number in front of the chemical symbol, as in table 3.1, has also been used in table 6.2. It is the fusion of hydrogen into helium which is the main energy source of the sun and other stars.

Two important points emerge. (i) The wavefunction, the crowning new concept invented for quantum mechanics, while representing the state of a system, is merely a calculational device. (ii) The results of quantum calculations are represented by probabilities. I shall now deal with this second point.

The 19th century, enthralled by Newtonian physics and its precise predictions of eclipses (for example), led to a belief in determinism. This means, in current usage, that events are determined by prior events. Even the fact that totally exact measurements do not exist does not upset this view, since the inexactness of measurements is merely a human failing. Even if the motion is chaotic (sections 5.1, 5.2), we still have deterministic chaos. The Newtonian notion of determinism is thus seen to cover a wider field today than it did in Newton's time.

Because quantum mechanics offers 'only' probabilities for the outcomes of experiments, it is profoundly different from classical mechanics. It undermines determinism in the following sense. The theory is probabilistic, i.e. not deterministic, but the probabilities evolve deterministically (via the *Schrödinger equation*). If probability arguments are essential to a physical theory, that theory can of course no longer be regarded as deterministic: the initial cause has a number of different consequences and with each of these we can associate a probability, so that we cannot be *sure* of the result. (Forty years ago a different view was still widely held [6.8]:

> *'Thus determinism lapses completely into indeterminism as soon as the slightest inaccuracy in the data on velocity is permitted.'*)

Things actually happen in this world, and after they have happened, everything is clear and complete. If we have a basic theory which yields only probabilities, we can ask: does there exist a better mechanics, call it an X mechanics, which assigns certainties? Is quantum mechanics in fact only a statistical version of an as yet undiscovered X mechanics? It could be, after all, like a form of a statistical mechanics which is based on classical mechanics. This idea works as follows.

If you go to an examination you may be told that the chance of any candidate passing is 0.7. If you study the past history and abilities of each candidate separately, you may be able to say: this candidate will pass, this candidate will not pass, and so on. This would correspond to a definite statement in some new 'X mechanics'. In the end of course 70% do pass. This corresponds to the idea that quantum mechanics is the statistical mechanical version of X mechanics. Quantum mechanics works and appears to be correct, but it does not go as far as X mechanics would go.

It is this search for an X mechanics, the elimination of probabilities at the basis of physics, that inspired Einstein's well-known opinion that God does not throw dice. Einstein was in search of certainty, but quantum mechanics did not seem to supply it.

This leaves what was regarded by some, and certainly by Einstein, as an imperfection at the basis of quantum mechanics. Instead of saying that this particle of this energy will be able to tunnel through this barrier, we can say merely that 'of 100 particles of this type, 20 will get through'. A test of this statement is of course to perform the experiment with many particles. This viewpoint leads to the notion that quantum mechanics can always be understood in terms of measurements on copies or 'collections' or 'ensembles' of systems [6.9]. These copies do not interact with each other, but can be thought of as in the same conceptual space. But they must differ in some detail which was left unspecified in the original definition of the system, for example in the way they are started off (initial boundary condition). It is only in this way that the ensemble interpretation suggests that quantum mechanics is in some sense incomplete (Einstein's view).

6.6   Attempts to understand quantum mechanics

'Hard-nosed' physicists include many who work on the structure of the nucleus and on elementary particles, those who try to bring about nuclear fusion for the production of electricity, people working on semiconductors, lasers and quantum optics. To them quantum mechanics is a beautiful and reliable theoretical tool, and that is that. The variables which occur are then accepted *not* to have their values until they are actually observed, in agreement with the Wheeler dictum, cited on p 125. In that sense the orthodox interpretation of quantum mechanics differs very significantly from the usual interpretation of physical phenomena.

However, the human mind wants more! There has been an attempt to really understand the individual processes in which electrons, photons and other particles participate, not *en masse*, but on a particle-by-particle basis, corresponding to the study of the individuals in the class, as described above. This reasonable curiosity has gathered momentum in the last 60 years, and has led to a great variety of views, embodied in an enormous bulk of publications, extending from [6.10] to [6.11].

Let us recognize that we have here an area of science in which apparent imperfection or incompleteness relative to classical ideas is pretty obvious. We should note Einstein's feeling that he did not really understand the nature of photons as well as Richard Feynman's view, echoed by others [6.12], that no-one really understands quantum mechanics. This is, however, a somewhat old-fashioned viewpoint [6.13] which we do not wish to emphasize here. Instead let us observe [6.14]:

> 'Probably the best way to agitate a group of jaded but philo-sophically inclined physicists is to buy them a bottle of wine and mention interpretations of quantum mechanics. It is like opening a Pandora's box...I have been amused to discover that the number of viewpoints often exceeds the number of participants.'

I shall try to convey the flavour of this work of interpretation.

6.6.1 Hidden variables

How would it be if there was really an X mechanics? You could argue that there are hidden variables which yield quantum mechanical probabilities, so that all is really known, only we are unable to determine the precise values of these hidden variables. If particles are thrown at the wall of a box, we see that some will enter the box by the tunnel effect and others will not. This seems to be a random affair, governed by probabilities. But if each particle has a certain 'hook' which determines if it will get in or not, then this is a 'hidden variable'. We do not know how to study it at present, but we would get rid of probabilities within the X mechanics. In this sense determinism would be back in physics. Note that the hidden variable concept is another invention connected with the development of the quantum theory.

The wavefunction would not now offer the best or most complete description of a system. The normal quantum states which have been discussed in previous sections would in fact be revealed to be averages over the hidden variables. To be acceptable, such theories must of course lead to experimental results which are in agreement with those known from normal quantum theory. So an important question arises: are hidden variable theories really possible? John von Neumann, famous co-inventor of the electronic computer, thought that he had proved that they were not (1932). But the proof was believed to be in error 34 years later (by John Bell (1928–1990)).

This reminds one of an error of his, pointed out 23 years later, in the complicated 1929 proof of a theorem (discussed on p 60). These two occurrences lead us to an obvious, but often neglected point: complicated mathematical arguments in physics always benefit from checks by intuition.

So we *can* have some form of hidden variable interpretation of quantum mechanics, and David Bohm (1917–1994) produced such a theory in 1952. It relied on a new proposed force, the so-called quantum force; there is no need for probabilities and particles always have definite positions and velocities. In fact, these *are* the hidden variables, and because of their occurrence the wavefunction no longer gives a complete description of the system. But there are snags: the exchange of messenger particles, normally pictured as responsible for a force (p 55), does not seem to occur here. Most pictures and theories in physics are *local*, meaning that events at a point in space and time are influenced by events only in its vicinity, and not by action requiring faster-than-light signals. The Bohm theory is non-local.

In it a particle rides on a 'pilot' wave. In the two-slit interference experiment (see p 132) each particle goes through one slit, but you do not know which, since the pilot wave goes through both slits. Further, the hidden variables are altered by measurements. An interaction between the particles via a pilot wave guides the particles; the pilot wave as well as the hidden variables are affected by the act of making a measurement. In this interpretation you find exactly the same experimental results as in the orthodox version of the quantum theory. Here determinism is restored at the price of introducing non-locality.

6.6.2 Non-locality

Locality is a term which expresses the idea that an influence that happens 'here' cannot instantaneously affect processes elsewhere. It needs some time to reach 'elsewhere'. Thus the energy of strongly ineracting quasi-particles cannot be associated with one location (p 64), nor can the photon energy (p 89), so that these are non-local quantities. The same applies to classical gravitational energy.

To see that this occurs more widely, consider two similar particles of opposite momenta, separating from each other, their total momentum being zero. If we measure the x component of the momentum of particle 2 at A, we can predict that of particle 1 at B from the momentum conservation law, although no measurement has been made on it. Alternatively, we can measure the x coordinate of particle 2 at B and deduce the x component of particle 1 at A, since the latter has moved symmetrically to particle 2. Thus particle 1 at A, though possibly far from the experimenter, snaps into a state of known momentum or into one of known position, depending on activities which take place far away from it, and it does so instantaneously. This is a non-local effect and valid in classical as well as quantum mechanics. In the latter it is more striking since position and momentum are incompatible, i.e. not measurable with precision at the same time.

What makes this thought experiment interesting in quantum mechanics is the probabilities of the various possible results of the measurement. If the particles at A and B are treated as independent entities, we obtain a result which is at variance with quantum mechanics. This is the subject of the 'EPR' experiment proposed by Einstein, Podolsky and Rosen [6.10] and elucidated later by many others (e.g. A Aspect, J S Bell, D Bohm). Quantum mechanics *is* regarded as valid, since the actual performance of this type of experiment has always favoured the results it predicts [6.15]. The particles can clearly not be regarded as independent, however far apart they are: they form a single system with an *entanglement* whose spatial extent is limited only by experimental constraints. It has no classical analogue. Although there are various interpretations and opinions on this matter, in that sense quantum mechanics can be regarded as non-local. This is still a matter for discussion [6.16].

It is the observations on particle 2 which bring into existence its properties, as well as those of the distant particle 1. Some people regard quantum theory as incomplete, as it cannot explain the properties of particle 1, although they are known in principle, since this particle does not interact with the apparatus.

Incidentally, any hidden variables would have to be *non-local*, since it was shown in 1964 by John Bell that *local* hidden variable theories of quantum mechanics are not possible.

6.6.3   Bell theorems

How does one show that in a certain situation hidden variables cannot occur? The basic idea can be gathered as follows. Suppose we are faced by three up-turned cups and two observers. They pick at random one of the three cups each and separate, possibly by a large distance. Upon a given signal they turn over their cups. One observer finds a black stone, the other finds a white stone. This happens again and again. Now let us try to explain this result in terms of the initial placement of black and white stones under the cups. If this succeeds we have found what corresponds to hidden variables, namely the initial placement of the available coloured stones. However, it proves to be impossible, since we have only black and white stones available, and there are three cups. So two stones are bound to be identical, suggesting that occasionally two stones of identical colour will be uncovered. But this is not found! The ingenious way out suggested by the nature of quantum mechanics is this: the stones are of indefinite colour initially, and colour is actually given to the stones by the act of turning up the cups, and if one becomes black, then the other has to become white. Thus locality has been lost in the initial placement of the stones.

The two stones correspond to two particles of opposite spins or of opposite momenta found by two observers in the EPR type experiment. My example is academic, but it is somewhat analogous to the details of the proofs given by Bell and the many authors who followed him in extending the breadth and depth of his theorem [6.16–6.18]. Again we see the point of the remark of John Wheeler's cited on p 125.

6.6.4 Schrödinger's cat

This legendary creature has been tortured for years. There are several books with this title and many papers on this subject. This 'experiment' was proposed by Schrödinger in 1935, and while the physics of the answer is accepted universally, no generally accepted interpretation of it has emerged. What is said below is a personal view.

The story is that there is a box containing a radioactive substance which can emit radiation in a direction such that it triggers a device which kills a cat which is also in the box. Assuming that the device and the cat can be described by quantum mechanics, the cat will after a certain time be in a superposition of an 'alive' and a 'dead' state. Yet when the box is opened the matter is decided: the cat is alive or it is dead. 'Explain!', as an examination paper would say.

An answer seems to be as follows. First point: suppose we have two possible future states A and B of a system, and one eventually turns up: e.g. a particle is in the box or outside it, as in the tunnel effect (section 6.5). To say that it is in some sense in both places is a misuse of language. There are probabilities for both situations, of which only one will turn up. But that is not all, and this leads us to the second point. You might suppose that the problem is similar to weather forecasting: it will rain or it will not rain this evening. So it would be a misuse of language to say *in the afternoon* that the evening weather exists in both forms! True; but, and this is the additional point, the two states in quantum mechanics are able to exhibit interference, and this makes the situation quite different from the case of the weather forecast.

A quite separate point is this: an enormous number of quantum states can be regarded as 'alive' and even more must be regarded as 'dead'. There will be many others to which these adjectives can barely be applied convincingly, and which must be considered to be partially dead, partially alive, etc. In fact the cat is really a 'quantum cat', a term I shall apply to these superpositions. They hold for a superposition of states for which we have no words, concepts or sense perceptions. They just cannot be appreciated in the macroscopic world. The observer cannot *see* a quantum cat. He can see only a cat, i.e. a normal terrestrial animal.

Particles collaborating in a single quantum state of a large system are said to be *coherent*. Such states are known to occur for large objects at low temperatures, for example when a metal can pass a current with zero electrical resistance (*superconductivity*). Similarly, some fluids can then move without internal friction (*superfluidity*). In order to exist such systems must avoid *decoherence*, which is caused by the many interactions to which a macroscopic object is subject. Think of all the photons and particles of the surroundings which interact with one surface layer of the body but not with a more distant surface layer, which is subject to different interactions. These many uncontrolled interactions destroy quantum superposition and tend to make our object a classical rather than a quantum mechanical one. Coherence 'leaks out', as it were, from quantum systems. This has been observed in subtly designed experiments (see for example, [6.19–6.20]). Decoherence is largely avoided for superconducting and superfluid systems by special precautions, particularly by very low temperatures. Such steps would also have to be taken to keep the quantum cat in its quantum state.

6.6.5   Entanglement

The two particles which are considered in the EPR experiments were once in each other's neighbourhood. However far apart they are in due course, you cannot remove this initial connection. Their states remain entangled with each other by virtue of the superposition of states (see p 137). Thus the temporal evolution of the state of one particle depends on the other.

If two particles do not move independently of each other, as in this case, their states are called *entangled*. (A slightly mathematical way of putting this is to say that the wavefunction for the two-particle state is then *not* a product of single-particle wavefunctions.) Of course, strictly speaking, everything is entangled with everything else in the universe. Note, however, that entanglement does not offer an opportunity for passing information between two bodies at a speed in excess of that of light. That is just as well, as relativity theory forbids this in any case. Independent particle motion occurs in the textbooks because it is easier for theoreticians to treat such situations, which approximate sometimes to the real (entangled) world.

A coin can be heads or tails, a spin up or down (or in a superposition thereof), a cat alive or dead. When we are told which it is, we impart a so-called *bit* of information. This is classical *information theory* language, as discussed on p 211. For two coins we have analogously four possible states. In quantum mechanics, we have instead a superposition of two states for one particle with two directions of spin. This represents a quantum bit, called a *qubit*. The four states occurring in the two-particle case are again in superposition and in quantum mechanics they represent two qubits. This is the newer *quantum information* theory language. It is worth noting here, as these superpositions open up the possibility of faster (quantum!) computations in the future [6.20]. A key problem is to guard against the destruction of the superposition of states involved as a result of the interactions with the surroundings.

6.6.6   The infamous boundary

If you conduct an experiment, you may have a complicated quantum system and have to extract pointer readings or observations from it. How is it to be done? It involves a boundary, that between object and apparatus [6.21], and characterizes the transition from the microscopic to the macroscopic. I showed above that this is a key aspect of the Schrödinger cat problem. There are many very clever experiments which have recently been done to elucidate these problems further. The upshot is that any attempt to change quantum mechanics, even by a slight amount, in order to render it more in agreement with so-called common sense, is liable to destroy the theory altogether, making it incompatible with the observed phenomena.

6.7   Comments on quantum mechanics

I now show that there are strongly held *different* points of view about the interpretation of quantum mechanics, by three citations, the last of which has my support.

(i) Well, if quantum physics is saying that a gun can be half fired and half not fired at a cat who is then half dead and half alive, or that the world contains a biological species that half exists and half doesn't, then this is just ridiculous. I am going to put this book down and forget all this nonsense! But it is the fact that these implications of quantum measurement theory are so absurd that is the main point of the argument. However successful quantum physics may have been in explaining the behaviour of atomic and subatomic systems, it should be clear by now that its statements about counters, cats and biological systems are quite wrong. What we hoped would be the final, fundamental theory of the physical universe is fatally flawed [6.22].

(ii) Concepts like 'measurement problems' and 'reduction of the wavefunction' refer not to what is happening physically, but to what is happening mentally. They refer not to the behaviour of physical systems, but to the way we learn about, interpret and understand this behaviour. They refer not to physics, not to properties of quantum mechanics, but to properties of our brains. They are not physics, but psychology—too often abnormal psychology [6.23].

(iii) Quantum mechanics can be understood....[those] writing for the general public do a disservice to science by clothing quantum mechanics in a mystical aura. It is a perfectly logical, coherent physical theory, which can be understood rationally. The mysticism is theirs [6.24].

Here is what a great physicist (Richard Feynman) is supposed to have said: 'Don't try to understand quantum mechanics or you will fall into a black hole and never be heard of again'. Nowadays we can do rather better than that!

6.8   Quantum effects

We now come to some important specific and unexpected effects which can be linked to quantum theory.

6.8.1 Black-body radiation

An *ideal gas* is defined to be one in which intermolecular forces are small, as they would be in the dilute gases introduced on p 11. While a theoretical fiction, it is a very useful one, particularly if you wish to make some simple calculations of gas properties. It will be understood that with millions and millions of particles in just one millimetre cube, if every particle feels a force due to every other particle, calculations regarding a *real gas* (i.e. one with reasonably strong inter-particle interactions) are impossible to do exactly.

Another important ideal system occurs if we construct a box with apparently nothing in it. It has to contain the radiation emitted by the sides of the box and if that is in equilibrium with the box, then, by the definition of thermal equilibrium, that radiation will be at the same temperature as the box. Scientists speak of radiation, or, which is the same thing, of light, as consisting of photons. These are packets of energy which can come in all colours. The radiation which is in equilibrium (for example with the container walls) is called *black-body radiation*. Note that appropriate mixing of light beams of different colours can yield a colourless beam of light. The converse is also true, as may be seen by passing white light through a glass prism selecting various coloured beams on the other side of the prism.

Imagine inspecting (i.e. 'scanning') the colours of the rainbow. Start on the side of the rainbow with relatively low energy (invisible) radiation beyond the red colour which has the self-explanatory name 'infra-red' radiation. Then pass from red to green to blue radiation. Finish eventually with the relatively high energy (invisible) radiation beyond the blue colour, called, again reasonably, 'ultraviolet', on the other side of the rainbow. Photons have all the colours of the rainbow, including colours which are invisible to the human eye. As each photon can for many purposes be regarded as a wave, it can be characterized by the number of oscillations which occur in a second. This is called the frequency (p 34) and in sound waves it corresponds to the pitch of a musical note. The energy of a photon is actually proportional to its frequency.

If we make a hole in the box, small enough not to disturb the radiation, we can investigate the photon energy distribution through this hole; for example we can determine how many photons there are in a given range of energy (or frequency), and do so for all energy (or frequency) ranges. This is a problem first studied at the end of the 19th century. The expression for the photon density for a given frequency range (the 'frequency distribution') has a beautiful simplicity if the box is large enough. It depends on just two variables: the frequency and the temperature, *not* on the shape of the box, the nature of the walls, etc. Thus, if we know of radiation that it is 'black body', then its (frequency or) energy distribution depends solely on its (absolute) temperature. We need not worry about small boxes, where the situation is more complicated [6.25].

To make sense of this distribution, Max Planck (see p 46) introduced the notion of graininess of energy: energy comes in the form of certain smallest lumps, which involve Planck's constant. The reader will recall that when you give energy to an atom, it cannot absorb less than a minimum which brings it to what is called its *first excited state*. I am not talking about increasing its energy of motion, but its own intrinsic energy. When it loses this energy by emitting a photon, it returns to its lowest energy, which is called its *ground state*. Thus Planck's discovery gave scientists an early clue to quantum theory, and he is here the reluctant *hero* of this chapter. I say 'reluctant' because he was a classical physicist at heart and did not really believe for a long time that quantum theory was as radical a departure from classical physics as it turned out to be (box 6.1).

Box 6.1 Max Planck 1858–1947: a short history.

23/4/1858	Born in Kiel, Germany.
1885–1889	Professor of theoretical physics in Kiel.
1889–1926	Professor at the University of Berlin.
1896	Wien's (1864–1928) empirical law for black-body radiation.
1900	Planck's quantum hypothesis for black-body radiation.
1905	W Nernst (1864–1941), who was Professor in Gottingen (1891–1905) becomes Professor in Berlin.

1909	Planck's first wife, Marie Merck, dies.
1910	Planck marries Marga von Hoesslin.
1913–1914	Planck becomes Rector of Berlin University.
1914	Einstein becomes Professor in Berlin, following a visit to Zurich by Planck and Nernst. Nobel Prize for Max von Laue (1879–1960).
1916	Planck's son Ludwig is killed at Verdun.
1917, 1919	His twin daughters both die in childbirth.
1918	Planck, aged 60 years, receives the Nobel prize for physics.
1919	M von Laue (1879–1960), Planck's student, becomes Professor in Berlin.
1920	Chemistry Nobel Prize for Walther Nernst.
1922	Physics Nobel Prize for Albert Einstein (in Berlin 1914–1933).
1928	Schrödinger becomes Professor in Berlin as Planck's successor.
1929	Awarded the Copley Medal of the Royal Society of London.
1944	The second son, Erwin, was until 30 January 1933 a Secretary in the Chancellor's office, and was involved in the anti-Hitler plot of 20/7/1944.
23/1/1945	Erwin is executed by the Gestapo.
6/5/1947	The talk between Planck and Hitler which took place in the spring of 1933 is the subject of a short note by Planck in *Physikalische Blätter* 3 1433 (1947).
4/10/1947	Death of Planck.

Returning to black-body radiation, we can now say: give me a black-body spectrum and I shall tell you the temperature of the emitting surface. For example, the sun's spectrum above the atmosphere, which is approximately that of a black body, is at 5760 K.

We have to look now for the incompleteness; things cannot really be as pretty as that. In fact, there are many examples of experiments yielding good approximations to the simple black-body spectrum described. The incompleteness resides here in the approximate nature of the agreement. However, we have also failed to consider that photons interact gravitationally with each other, so that we want to know the properties of the radiation, allowing for this effect. The results are then more complicated, see section 6.9 and [6.26].

6.8.2   Virtual pairs and the Casimir effect

Is energy really constant? Are there no exceptions? Is there no imperfection left? In fact quantum theory has given us the uncertainty principle, which implies a kind of intrinsic imperfection (using the term of Chapter 1). It dictates that, given a short time interval, the energy of a system is uncertain, and this uncertainty is greater the shorter the time interval. Nature is here a kind of bank manager: the more energy you want to borrow, the shorter the time for which she will allow you to have it. To this extent the classical notion of energy conservation actually fails. During these tiny periods of time particle–antiparticle pairs (see antimatter on p 50) may come into existence and then decay. They are called virtual particles or virtual pairs. This must strike the reader as a profound change of tune. Is there a chance of perpetual motion of the first kind after all? No, since these violations of energy conservation last only for tiny periods. The state of lowest energy of an empty box can thus be pictured as a seething mass of particle–antiparticle pairs which are created and get annihilated all the time.

An empty box? This sounds like what we used to call a vacuum. It is indeed still a vacuum, but it contains this great activity even at the lowest temperatures. The virtual particles are associated with waves (as are real particles). Suppose two parallel metal plates are inserted into the vacuum with a separation of only a few millionths of a metre. We can then measure an attractive force between them. The reason is that only some complete wave trains corresponding to virtual particles can be accommodated between the plates. There are no such constraints for the region outside the plates, where all wavelengths can be accommodated, and this lack of balance has the effect of a force which tends to push the plates together. It is as if the vacuum outside was banging at the plates asking to be allowed in. There are many relevant experiments, and they are not confined to metal plates nor to light. The effect also exists for sound waves [6.27]. It was predicted in 1948 by the well-known Dutch physicist H B G Casimir (1909–2000).

By allowing the Casimir force to compress a spring, we could use a pair of parallel plates to store energy and produce useful work. The vacuum would become an unlimited resource [6.28]! A fantastic idea, not currently accepted, with a yield of less than a millionth of a Watt—but can we rule it out? After all, Luigi Galvani's (1737–1798) work on the effect of electricity on frogs legs led eventually to commercial electricity; and there is an apocryphal story that when a high official asked Faraday (1791–1867) about 'the use' of his work on electricity, he is supposed to have replied: 'I do not know, Sir, but one day you may be able to tax it!' It is perhaps a bit too hopeful to suppose that history would repeat itself in this instance.

We can say that energy conservation holds only in the statistical sense, a view that was already championed in the 1920s [6.29]. We can discern a problem here. Take a detached view of a normal large system, and energy conservation holds. Look in microscopic detail, and the conservation law becomes less absolute and turns out to be only a statistical law.

6.8.3 Screening

The state of lowest energy of a system is called its ground state in quantum theory. If longer-lived particles, i.e. real (as contrasted with virtual) particles, are also in the enclosure, this jiggling going on between any pair of real particles, including photon emission and absorption, must be expected to cause a kind of *screening* (or weakening) of the forces between them. This is a highly simplified picture of a sophisticated effect. This screening, though very small, was studied experimentally in 1947 for low energy levels of the hydrogen atom by Willis Lamb (born 1913; NL 1955). A small effect was found and is now called the Lamb shift. It caused great excitement because everything to do with the hydrogen atom, which is the simplest known atom, is regarded as fundamental, and also because of the tremendous accuracy required in finding the Lamb shift.

Nothing is entirely new under the sun, however, and more conventional screening had been observed and investigated before

(P Debye and E Huckel 1923). But they dealt with an electrolyte, i.e. normal material, not a vacuum, and the intervening particles were electrons and ions, i.e. real charged particles. Various other types of screening became important later on. The reason is that the screening concept provides an approximate way of thinking about the tremendously complicated problem involved in the interactions of millions and millions of particles. This idea has also been used extensively in the study of electrons in solids.

6.8.4   Particles at rest?

Can a particle actually ever *be* at rest? The answer is 'no', and it applies with greater force the smaller the mass of the particle considered. This may again be thought of as a result of the zero-point motion (see p 12).

Alternatively, it can be thought of as due to the uncertainty that the position and the speed of a particle cannot be known precisely at the same time. Thus a particle in a box is known to be somewhere within it; so it must have an uncertain speed or momentum, and is therefore not at rest.

To find the atom which has been most *nearly* at rest in experiments, we have to go to quite recent work. A beryllium atom is confined to a small box of linear size 1 millimetre. It has an average velocity calculated to be about 1 centimetre per second. This is very small indeed for an atom, as can be seen by estimating its equivalent temperature. This is a temperature which corresponds in some clear-cut manner to the average over all molecular speeds in a system: the faster they move, the higher the equivalent temperature. For our beryllium atom this temperature is just a tiny one ten millionth of a degree Kelvin [6.30]. We see: 'In nature nothing remains constant. Everything is in a perpetual state of transformation, motion and change' [6.31]. For normal objects this constant fluctuation effect is present, but usually negligible. Photographs of atoms in an appropriate trap were achieved by Hans Dehmelt (born 1922; NL 1989) and his colleagues in the 1980s.

6.8.5 The *Unruh* effect

The above description of a vacuum is subject to a constraint which all 'normal' people accept, namely that it is observed by somebody not acted upon by outside forces, i.e. they are sitting at rest within it, or, possibly moving with constant velocity through it. Such people are called 'inertial' observers. They are the key actors in any exposition of special relativity theory. But we are *not* 'normal', since we are interested in the nooks and crannies of the world around us. So: suppose the observer is subject to a constant acceleration through the vacuum. Then the particle–antiparticle creation and decay appears to be split up in some way, and he sees an enclosure of photons apparently in equilibrium with themselves, at a temperature which turns out to be proportional to this acceleration [6.32]. We can appreciate that a new (*'Unruh'*) temperature will be observed, since the radiation modes of the vacuum will appear accelerated to our observer, and will therefore not correspond to the true vacuum. However, this effect is numerically almost irrelevant: an acceleration of one hundred million million million times the acceleration due to gravity (on the earth) is needed to give a temperature rise of one degree Kelvin!

The vacuum is a volume of space cleared of matter, but it is subject to electromagnetic fields and their irreducible fluctuations, particle–antiparticle creation and annihilation, radiation pressure and the like. If the effect of gravitation is added one arrives at a complicated physical entity which is in fact not fully understood [6.33].

6.8.6 Measurement without interaction?

This curious effect can be demonstrated with the well-known instrument shown in figure 6.10, called a Mach–Zehnder† interferometer. It guides photons along two distinct paths by splitting the incident beam at B_1, and using two mirrors M_1, M_2. The path can be arranged so that only detector D_1, but not D_2, is activated. This is due to the interference of waves i and j, and the symmetry of the two paths shown. If an object O absorbs i, this symmetry is destroyed, and D_2 would register as well. In this sense the photons j serve to show the presence of O without having interacted with O.

---

† Ludwig Mach, not Ernst Mach.

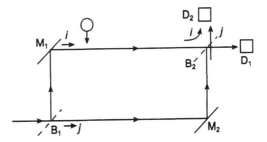

Figure 6.10 A Mach–Zehnder interferometer.

A Elitzur and L Vaidman, who found this effect in 1992, tell the story of a pile of bombs being delivered with the request to find the dud ones—a hard and dangerous task at the best of times. It is stipulated that a bomb explodes if a photon hits the movable mirror at its top. For dud bombs it is rigidly attached. One solution of this problem is to make the bomb mirror the mirror M_2 in the figure. It is then dud if no signal is registered by D_2. Of course you lose bombs which explode during this rather theoretical 'experiment'.

6.9 Can gravity affect temperature or light?

6.9.1 Mass is energy

We have already seen on p 21 that mass is equivalent to energy. Actually a particle in motion has, as you might expect, more energy (as well as momentum) than it has when it is at rest. Its so-called kinetic energy is just the difference between the two energies.

The energy locked up in matter is enormous. If a sack of 139 kilogram of matter (or two fully grown men) gave up all its mass as energy, one would have enough energy (3.5 million million kilowatt hours) to supply the whole of the USA with electricity for a year at 1994 values (production of electricity: 3.47 million million kilowatt-hours) [6.34].

6.9.2 Conservation laws

A second consequence of relativity theory was a unification of the conservation laws of mechanics. The laws of conservation of momentum, mass and energy all appeared in the period from 1600 to 1900, and then the 20th century wrapped them all up into a single law. This kind of unification of theories and ideas is characteristic of the drive for better understanding, but also for elegance and even artistry in theoretical physics. We come across it again and again. There is, according to special relativity, one basic entity, which is specified by four numbers, and which is subject to a conservation law. It is made up of the three space components of momentum, and one additional component which represents energy, and so includes the relativistic energy due to the mass which the particle has at rest. Because it is conserved, scientists now need to consider the conservation of only this single four-component entity, whose individual components (e.g. the classical mass) need not be conserved in relativity. Its name is the *energy–momentum four-vector*. This idea is not all that hard to understand: a traveller may have a reserve of dollars, pounds, francs and lire. This 'reserve' is specified by four numbers; the currencies may be interchanged as she travels, but she keeps the total fixed. Very roughly this reserve corresponds to the conserved quantity. (Actually, the 20th century has also given rise to new conservation laws.)

Tests of conservation laws also come from the observation of collisions between elementary particles (discussed in section 3.9). These particles leave tracks in bubble chambers or on photographic plates. Conclusions about energy and momentum can be inferred from the various characteristics of these tracks: their lengths, directions, and so on. The reason is that each particle brings to a collision energy and momentum, and the final products also have energy and momentum. So you can paraphrase for collisions what Lavoisier said about the masses in chemical reactions: the total energy and the total momentum must be the same before and after a collision.

6.9.3 Gravitation and photon paths

Let me leave this matter of unifying conservation laws, and return for a moment to an exciting consequence of relativity, namely that energy has mass associated with it. It means that a body which absorbs radiation, typically the rays from the sun, gains mass energy. All bodies radiate energy; indeed, the hotter they are the more energy they radiate. So it is no surprise that a radiating body loses mass. One can deduce from this idea two very interesting results which were not fully accepted before the 20th century. They are that gravitational effects change the path of a light ray and also change its energy. Speaking roughly, light energy travels in energy packets called photons (p 41), so that we are discussing how gravitation affects both the path and the energy of a photon.

We all know that matter attracts matter by gravitation—remember Newton's apple! Since light carries energy, and is therefore associated with mass, it follows that light should be attracted by matter. So a ray of light, which would otherwise travel in a straight line, will be attracted by a nearby star to trace out a curve which is bent towards the star. The deflecting body must be very large for this effect to be observable. For example, the sun deflects light from distant stars. This can be observed well during a solar eclipse, as this darkens the whole sky by also removing much reflected solar light, thus enhancing the visibility of starlight. This deflection was found by the British eclipse expedition of 1919, and confirmed in later observations. The order of magnitude of the effect was explained numerically by Einstein's general theory of relativity (1915). Newton's old corpuscular theory of light predicted a similar effect; but it was too small (by a factor of about 2).

6.9.4 The gravitational red shift

The second effect of the gravitational pull on light—the change in energy of a photon—can be explained by considering a perpetual motion machine of the first kind. It has already been seen that such machines cannot exist, because, by energy conservation, energy output equals energy input. Still, it is sometimes useful to think of such a machine even though you would never dream of rigging one up. Then you can say: 'Ah, I have a perpetual motion machine! So there is something wrong'. When you have found what is wrong, you have (perhaps) discovered a new effect. This is a kind of *reductio ad absurdum* argument, in which you show that *without* the new effect you arrive at a nonsensical situation. We will now pursue this kind of argument.

Our perpetual motion engine consists of two identical atoms, connected by an 'ideal' (e.g. weightless) string which passes over an ideal (i.e. frictionless) pulley. That's all. The idea is that the atoms exchange photons of just the correct energy to put an atom into its first excited state on absorption and, on emission, it drops back into its ground state. By proper optical channelling the energy is assumed to travel just between the atoms without involving other matter. The higher atom absorbs the photon, and so becomes heavier, and descends. Near the bottom the atom is arranged to re-emit the radiation so that it is absorbed by the higher atom. Now it (the higher atom) has become heavier, and so *it* goes down. Near the bottom it emits the radiation again, and so on, and so on. The set-up is like that envisaged in the 'bricklayer sketch', where a bucket full of bricks is connected to the bricklayer over a pulley. The bucket crashes down from the roof, so that the bricklayer is dragged upwards, and hits the bucket at mid-point as he ascends. The bucket breaks, the bricks spill out, and the bricklayer, being now heavier than the bucket, crashes down hitting the remnants of the bucket half-way up. Of course this sorry tale is much funnier when related by Gerard Hoffnung!

Returning to the case in point, this atomic see-saw could go on indefinitely, and the pulley might work a clock (say). It is in fact a *perpetuum mobile*. So there must be something wrong. The machine has to fail! We can infer here a new effect: as a photon rises against gravity it must lose energy so that the 'weaker' photon is unable to raise the upper atom from its lowest ('ground') energy state to its next higher energy. The see-saw is then impossible. We see that relativity predicts that light energy is affected by gravity. In fact, any photon loses energy as it rises against gravity. This effect was confirmed in 1960 by R V Pound (b 1919) and G A Rebka in experiments at Harvard University and it was also confirmed at Harwell, UK [6.35].

So how shall we describe the energy (or frequency) loss of a photon as it rises in a gravitational field? It is of course a 'gravitational *red* shift'. This is a fine, frequently used, term. The effect was predicted by relativity theory. This confirmation caused considerable excitement because of the high accuracy involved. It is like measuring the distance from here to Venus with an error of less than 1 cm; or, roughly, as if you wanted to measure a time interval of 100 000 years to within a second or so! It is this tremendous accuracy which captured everyone's attention.

Einstein had died five years before this particular confirmation of the theory of relativity. Actually, as far as he was concerned, these problems (of relativity theory) were long ago clear to him, and experimental confirmation would not have excited him unduly. Indeed, when one of his students expressed joy about the positive results of the British eclipse expedition in agreement with the relativity prediction, he is reported to have said 'But I knew that the theory is correct'. When asked what he would have said if there had been no confirmation, he is reported to have said 'Then I would have been sorry for the Lord—the theory is correct' [6.36]. The gravitational properties of the photon are encapsulated in a little parody (based on the Rubaiyat of Omar Kayyam) due to Eddington. The last verse reads

> *Oh leave the Wise our measures to collate.*
> *One thing at least is certain,* light *has* weight.
> *One thing is certain, and the rest debate—*
> *Light-rays, when near the Sun,* do not go straight.

6.9.5   The photon as a particle

We have seen (see p 136) that photons can behave as particles. This picture was used by Einstein in 1905 to explain the fact that photons hitting a metal surface have to have a minimum frequency (i.e. a minimum 'quantum' of energy) in order to liberate electrons from that surface. He did not use the term 'photon', of course, which was proposed only in 1926. Surprisingly, the ability to liberate electrons does not depend on the number of photons in the incident radiation, but only on the energy per photon, i.e. on the photon frequency. Photons are seen to be packets of energy. Thus this so-called *photoelectric effect* could be regarded as due to collisons between photonic energy packets and the electrons in the solid surface.

When I speak of a 'particle' or 'elementary particle', I always mean a quantum particle, i.e. one which can exhibit wave or particle properties, depending on the experiment to be performed. Furthermore, if the particles have the same characteristics such as mass, charge, etc, then they are regarded as strictly *indistinguishable*. This is reasonable since there is then no experimental technique which can distinguish between particles. Let me now return to our main theme.

6.9.6 Gravitational slowing of clocks

There is a further striking implication. We saw earlier in this section that the photon energy and frequency are lowered by letting the photon rise against gravity. It follows that a larger number of cycles at a higher level in a gravitational field, say on top of a mountain where the gravitational field is weaker, pass a fixed point than would be the case at the bottom, where we as observers are sitting. This means that a clock at the top of the mountain appears to go faster than the same clock does at the bottom. Gravity is stronger at the bottom, closer to the earth's surface, so we have another surprise: the gravitational slowing of clocks. It applies to any clock, independent of its structure. It is a relativistic effect closely connected with the gravitational red shift. It was confirmed experimentally in 1980 [6.37] by flying a maser clock in a rocket and comparing it by radio with a clock on the ground.

6.9.7 Gravitation and thermal equilibrium

We might suppose that in an equilibrium state the temperature of a system is uniform throughout. But we did not define an equilibrium state in this way (section 2.4). Why? Because you might ask: is it strictly and universally true that in equilibrium a system is at a uniform temperature? The answer is 'no'. Let us look at it this way: on earth we are in a gravitational field, this is a another way of saying that things drop to the floor when we let go. In fact nothing can escape the pull of gravity, not energy or even light. In an analogous way we find that in a gravitational field heat, being a form of energy, also tends to diffuse, or 'sediment', to the bottom. It 'sediments' like sand does when it is carried in river water. Therefore we would expect a column of gas in a gravitational field, even though it is in thermal equilibrium, to be slightly hotter at the bottom of the column than it is at the top.

This is a surprise! It is clearly unwise to define thermal equilibrium, as is often done, as a state of uniform temperature. If you omit the term 'thermal' and talks about equilibrium, pure and simple, then experts would include the condition that the pressure must also be uniform throughout the column, and this is violated in a gravitational field, since the gas pressure is greatest at the bottom of the column. The gas column in a gravitational field would then *not* be considered to be in equilibrium (omitting the word 'thermal'). Thus we come back to an equilibrium state as one in which no more systematic changes occur if the system is isolated.

6.10   Matter drained of heat

The atoms and molecules which are moving about in all our systems move less and less as the temperature is lowered. As we have already noted, gases tend eventually to become solids. The free motion of the atoms is then reduced to oscillations of these atoms, which are now more or less tied to fixed points in the solid. If the solid is crystalline, then the atoms are arranged in a three-dimensional grid, called a *lattice*, and each atom is vibrating about its home point in the lattice. Does this jiggling stop as the absolute zero of temperature is approached? The answer is that it does not. For, if it did, an atom would have a definite position and speed (namely zero speed), and this contradicts the requirements of the uncertainty principle. Hence we are again left with zero-point motion. It is particularly pronounced in the case of normal helium gas. Because of its high zero-point energy it becomes liquid only at low temperatures (4.2 K for normal atmospheric pressure) and is in fact the only element that can continue to exist in the liquid state near 0 K. It was liquefied only in 1908, much later than the other *noble* gases. From its normal liquid state, known as He I, it suffers a transition to the amazing He II at 2.17 K. The latter contains a *superfluid* component by virtue of which the liquid creeps out of any vessel by spreading over the container wall. There is practically no friction to impede its movement. An early explanation of this effect depends on the following idea.

Imagine a ladder whose rungs represent energies: the higher rungs represent higher energies. Most enclosed systems are associated with such an energy ladder. The simplest 'systems' are particles which can move up and down their ladders. If the larger system (in which these particles are) gets hotter, there is a tendency for the particles to climb up the ladder. They tend to drop down if the system is cooled, but there is always a distribution of particles on the rungs. The particles are related in the sense that if two fermions (see p 47) manage to get hooked to each other, they become a boson (see p 47) and so transfer to a boson ladder. They are called *Cooper* pairs after L N Cooper (b 1930), who with J Bardeen (1908–1987) and J R Schrieffer (b 1931) obtained the 1972 Nobel prize for their theory of superconductivity. But three fermions locked together are again a fermion, and so on. The normal helium atom consists of six fermions: two protons, two neutrons and two orbital electrons and so is a boson. The more usual nomenclature is to call the rungs of the ladder 'quantum states'. If you want to emphasize the relative heights of the rungs, it is more convenient to call them 'energy levels'.

As the temperature of a system, be it of fermions or of bosons, is lowered, the lower energy quantum states become more crowded at the expense of the upper states. The lower temperature leads to the system becoming less energetic. It is like standing, sitting or lying down: the weaker one is, the lower one's energy level! So, as part of the pursuit of the elusive absolute zero of temperature, bosons will crowd into the lowest energy quantum state, a phenomenon called Bose–Einstein condensation (BEC). This is a phase transition, as discussed in section 5.4, characterized by a sudden change of some variable, such as the heat capacity, with temperature. Theory suggests that this would be impossible for a system approximating a one-dimensional line.

It is thought that superfluidity can be attributed to BEC (Fritz London 1938). So superfluidity was not then expected in a gas of fermions such as a certain variety of helium, called helium III, whose nuclei consist of two protons and one neutron and which has two electrons orbiting its nucleus. However, it was found in 1971 that these helium III atoms can pair up, making it a boson gas after all. Its temperature would be below a three-thousandth of a degree Kelvin. It can then exhibit BEC, as evidenced by the occurrence of superfluidity. The 1996 Nobel prize in Physics was awarded to three Americans for some of this work. There has been much recent activity in trying to find BEC in other two-fermion systems.

In current research Bose condensates are created at one ten millionth of a degree Kelvin—leaving, of course, absolute zero still unattainable. We are able to create condensates with an increasing number of atoms: 500 000 sodium atoms in 1995, a million atoms no doubt soon. The condensates are now big enough to be photographed [6.38]. The atoms are linked so effectively that we can think of such a system as a single *superatom*. Drops of these Bose–Einstein condensed superatoms may serve as a new type of *laser*, namely one which uses atoms rather than photons [6.39].

Because of their size and their visibility, these superatoms cannot be regarded as belonging to one of the three normal states of matter, namely solids, liquids and gases. Nor are they in the *plasma* phase (which accounts for 99% of the contents of the universe), in which the kinetic energy of motion of the atoms is sufficient to cause considerable rupturing of atoms so that you are left with electrically interacting atomic remnants (ions and electrons). They therefore qualify to be called an 'additional' state of matter. In fact a quark–gluon plasma (see p 54) was created experimentally in recent years. Other states of matter which have been added fairly recently include amorphous materials whose atoms do not sit on lattices.

6.11   A look at superconductivity

This phenomenon, already mentioned on p 146 and discovered in 1911, refers to the perfect electrical conductivity of some materials below some low critical temperature T_c. After decades of attempts, in the 1950s, the physical origin of this phenomenon was understood. We were able to increase T_c from 4 K (for mercury, by Kamerlingh Onnes) to 23 K for a germanium compound in 1973. This 'classical' period of superconductivity was replaced by a new field, *high temperature superconductivity* (HTSC), when a T_c of 35 K was found in 1986. Later higher T_c values were reached for compounds of, typically, barium, copper and oxygen (figure 6.11).

The technical implications are considerable as it is cheaper and easier to cool by use of liquid nitrogen at 77 K than by liquid helium. This made possible the wider use of superconducting magnets to cause levitation of fast-moving trains such as the Magler train in Japan, the superconducting wiring of electric motors, magnetic resonance imaging (MRI) for medical purposes and SQUIDS (superconducting quantum interference devices), for example as sensors for foetal heart beats. Figure 6.11 shows how the T_c value has risen in recent years.

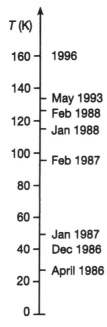

Figure 6.11 Highest T_c values reached recently.

As HTSC is another so far imperfectly understood topic, I shall just explain roughly how classical superconductivity arises. The long-range electrical (or *Coulomb*) interaction between electrons is reduced (screened, p 153) to short range by the background of the positive ions and the many electrons. Since the electrons attract positive ions which themselves attract electrons, this leads to an electron–lattice interaction. This may be pictured as an exchange of phonons between electrons which can then form Cooper pairs (p 163), and this leads to superconductivity. The attractive interaction depends on the mass of the ion involved, which is not fixed for a given material because of the existence of isotopes (p 32), and is expected to decrease as the isotopic mass is raised. This observed *isotope effect* of superconductivity [6.40] provided an early clue to the previously unexpected relevance of the electron–lattice interaction to classical superconductivity. Herbert Fröhlich, who made this crucial observation, missed the 1972 Nobel prize.

6.12 Summary

In this chapter I outlined the intellectual journey which led man to associate waves and particles with the same object. It is reminiscent of the possibility that one and the same drawing can represent a young woman and an elderly woman at the same time (figure 6.12). Ambiguity can be stimulating, and it occurs in other contexts: in art (in anamorphism and in Escher's drawings) and also in literature.

Figure 6.12 In 1925 caricaturist W E Hill represented a young and an elderly woman in the same picture. The chin of the former is the nose of the latter. The young woman's left eye is the elderly woman's right eye. Could this be analogous to the wave and particle aspects of electrons?

We know that the energy of a body can in Newtonian mechanics vary from some small number to some upper limit. How, then, can you represent a bunch of numbers like the energy levels of a hydrogen atom by a single quantity and call it its energy? You can see at once the need to invent new concepts (p 126). From an analogy between mechanics and optics you arrive next (p 127) at the concept of matter waves and Bohr orbits. This is part of a new mechanics, which requires the introduction of yet more new concepts, notably the idea of a wavefunction (p 137). Thereafter we had to grasp that quantum mechanics is a probabilistic theory, and I discussed this feature. There are several interpretational schemes for the new mechanics (p 148ff).

In sections 6.8 to 6.11 I have bombarded the reader with several current ideas. However, it has to be admitted that there are many more which I have not mentioned. The difficulty of specifying position and speed of a particle simultaneously is another limit to knowledge and it is used to good effect on p 152 (to query energy conservation and to suggest zero-point motion). A different incompleteness, which is not intrinsic but unjustifiable, often arises from the neglect of the effect of gravity on both light and temperature (p 156). It corresponds to thinking of our systems in the zero-gravity limit, and is therefore a 'limit imperfection'. I have also drawn attention to the astonishing properties of matter found in pursuit of the elusive absolute zero of temperature (p 162).

Returning to p 157, the conservation laws emerge rather surprisingly from what we can call 'continuous symmetries'. For example, time-reversal symmetry T (p 75) can be shown to lead to energy conservation, while symmetry under reversal of a spatial direction leads to momentum conservation. This is part of a result established in 1918 by Emmy Noether (1882–1935) of Göttingen—a great mathematical genius. Her theorem affects the structure of many interactions in physics.

It is reasonable to pass now from the very small to the very large, which we will do next.

Chapter 7

The galactic highway
Cosmology: science as history

7.1 Ages

Box 7.1 Age of the earth.

'But I desire to point out that this seems to me one of many cases in which the admitted accuracy of mathematical processes is allowed to throw a wholly inadmissible appearance of authority over the results obtained by them. Mathematics may be compared to a mill of exquisite workmanship, which grinds you stuff of any degree of fineness; but nevertheless what you get out depends on what you put in; and as the grandest mill in the world will not extract wheat-flour from peascods, so pages of formulae will not get a definite result out of loose data.'

You can almost hear the precise voice of a man of substance, uttering a warning which is not out of place even today, almost 130 years later. The place was the Geological Society of London, the date 19 February 1869, and the man the great evolutionist T H Huxley. He was speaking in favour of taking the age of the earth as about one hundred million years. This was in agreement with the order of magnitude required by Charles Darwin in his *Origin of Species* (1859). So why argue?

They had a powerful opponent in William Thomson (Lord Kelvin, see section 2.1). He had studied the recently developed Fourier theory of heat conduction and had applied it to the cooling of the earth from its primordial state. He assumed that the heat of the earth came from the sun, giving the original molten state of the earth and added heating of the earth due to gravitational contraction. He arrived at an age of the order of 20 to 40 million years. He revised this estimate from time to time. However, much longer time spans were needed to account for certain features of the earth's crust, notably the removal of solid material from the chalk cliffs of Kent by water. Thomson had some support from other physicists, and it seemed that naturalists and physicists were in opposition. What to do?

When two opposing views like this are held for a while, the chances are that in a sense both are right. Thomson had remarked that his estimates assumed that no new source of heat was available. However, radioactivity was discovered near the turn of the century. It was realized that enormous energy was stored in radioactive substances, and seen how elements could be 'transmuted' (see section 3.9) into entirely different ones. It gave birth to new methods of *radioactive dating*. In this method we find the rate of transformation of a radioactive element into a final stable product. The law of decay is characterized by a certain time (called the *half life*). Knowledge of this time and of the ratio of the masses of the initial and final products then yield the age of the mineral in which they are found.

Radioactive decay within the earth added a source of energy so as to increase Thomson's estimate of the age of the earth. This meant that the calculations made by Lord Kelvin (as he became later), while sound in themselves, were not based on correct assumptions. It was Ernest Rutherford (1871–1937; NL in Chemistry 1908) who made the connection. He discussed this at an invited lecture at the Royal Institution in 1904. Here is his amusing recollection of the occasion [7.1]. For more background, see [7.2, 7.3].

'I came into the room which was half dark, and presently spotted Lord Kelvin in the audience and realized that I was in for trouble at the last part of the speech dealing with the age of the earth, where my views conflicted with his. To my relief Kelvin fell fast asleep, but as I came to the important point, I saw the old bird sit up, open an eye and cock a baleful glance at me! Then a sudden inspiration came, and I said Lord Kelvin had limited the age of the earth, provided no new source of heat was discovered. That prophetic utterance refers to what we are now considering tonight, radium! Behold! the old boy beamed upon me.'

Today we think of 4.5 thousand million years as representing the *age of the earth* as well as the ages of the oldest rocks. Unicellular life in its earliest forms is about 80% of this age. Fossils go back about 700 million years and mammals 150 million years.

While discussing ages, let us throw in the *age of the universe* itself. According to current ideas it is about 13 thousand million years old. Like the other ages, this is only approximate and liable to be changed from time to time. However, the order of magnitude, which alone concerns us here, is probably sound. Some scientists (notably Sir Fred Hoyle) believe that the evidence suggests a mere 8 thousand million years for the age of the universe and this generates problems with certain stars, which are then older than the universe! We need not concern ourselves with this problem here, except to note that uncertainty arises from time to time regarding the age of the universe itself.

The universe was not always believed to be that venerable. Bishop J Ussher (1581–1656) claimed to have shown from the ages of the patriarchs as recorded in the Bible that the universe was created on Sunday 23 October 4004 BC! This time is represented by the lowest point in figure 7.1. The estimates increase the age as time marched on. But between the years 1500 AD and 2000 AD the estimate is seen to have increased by far more than just 500 years (figure 7.1)!

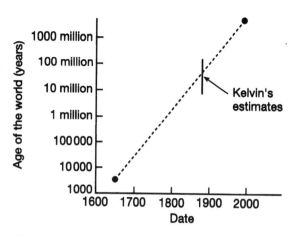

Figure 7.1 The 'age' of the universe through the ages. The vertical line represents some of Lord Kelvin's estimates.

7.2 Hubble's law

When a train moves away from you the pitch of the guard's whistle seems lower than it is when the train is at rest. When the train approaches the pitch is higher. This common effect is associated with the name of the Danish physicist Christian Doppler (1803–1853). He pointed out that there appears to be a compression of the sound wave when it travels towards the observer. This shortens the wavelength (see p 126) and so raises the pitch. Conversely, it is stretched when the sound source moves away, and the pitch is lowered (*Doppler effect*). As in section 6.3, you can transfer the qualitative effects for sound to optical phenomena: you expect a blue shift for an approaching light and now a new red shift for a departing source.

In the astrophysical context these shifts can be considerable because the velocities involved are large. Atoms can still be recognized by their 'fingerprints' (p 34), i.e. by their spectral lines, but these are now shifted bodily. The Doppler shifts can then be used to find their speed of approach, or, more usually, their recessional speed.

We often consider *galaxies*. These are, with an accuracy to within some factors of ten, systems of about one hundred thousand million stars loosely held together by gravitational forces. Our own galaxy, the Milky Way, was in due course judged to be 100 000 light years in diameter. Its thickness is about 2000 light years, and it contains about two hundred thousand million stars. A jet plane would take more than two thousand million years to cross it!

Once we know the speed of astronomical objects, it is highly desirable to know how far away they are from us. *Direct* distance measurements can only be made for closely neighbouring stars by noting how they appear to move around the sky as the earth moves round the sun. This (*parallax*) method is rather like noting from a train window that distant houses appear to move more slowly past the window than nearby trees. Parallax is also involved when an object appears to move slightly if you look at it first with one eye and then with the other. The parallax effect can be made a basis for a measurement of the distance of the houses from the railway line. Taking a *light year* as the distance which light can cover when moving in a straight line for one year, we can say that the parallax method works for stars which are 'only' up to about 100 light years away from us. Our nearest star is *Alpha Centauri* (4 light years).

For larger distances *cepheid variable stars* are essential. They attained importance when in 1912 Henrietta Swan Leavitt of Harvard College Observatory made an early systematic study of them. They are stars whose luminosity varies periodically. There is a reliable relation between the period of their luminosity variation and their *absolute* (or intrinsic) *luminosity*. Thus the latter can be inferred from the period, which is comparatively easy to obtain. The *apparent luminosity* on the other hand depends on the actual light energy received on earth. Comparison of these two luminosities yields the distance from us. This is a little like inferring the absolute luminosity of a light bulb, if you know its distance away and observe its apparent luminosity. The procedure for determining astronomical distances is really much more involved and still presents a major challenge for astronomers. In the 1950s an error was in fact found and the previously inferred distances had to be greatly increased.

By the methods described it fell to Edwin Hubble (1889–1953), an American lawyer who became a famous astronomer, to find in the 1920s that there is a tendency for stars, and galaxies of stars, to recede from us with a speed which is proportional to their distance from us. Hubble's work enabled us to plot the recession velocity of selected galaxies as a function of their distance from us. Each observed object furnished a point on the graph, and these points suggested a straight line, rather like figure 2.1, and shown in figure 7.2. There are many ways of plotting such graphs, and Hubble's actual graph of his famous 1929 paper is not given here. For example, you can plot an important parameter, the *red shifts*, of a certain class of astronomical objects. This is the wavelength shift, namely the observed wavelength reduced by what has been calculated to be the emitted wavelength, expressed as a fraction of this emitted wavelength. It is usually denoted by z. Many z values are quite small: four thousandths (for the Virgo cluster). But large ones $z \backsim 12$) are also known.

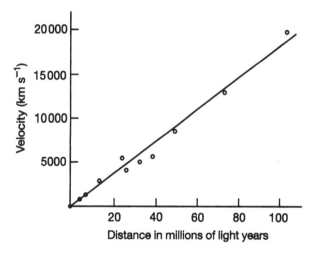

Figure 7.2 The velocity–distance relation. Circles: mean values for clusters of galaxies. Dots near origin: galaxies used in the original Hubble formulation [7.4].

We can guess that the rules for the flight paths of galaxies include some familiar ones: 'no U-turns, no overtaking'. It is easily verified that the Hubble law relative to one galaxy implies that it also holds relative to any other galaxy: galaxies separate from each other like crosses on an expanding balloon.

Some summarizing remarks about astronomical objects follow. They are of course bound to leave out many details and a vast amount of research. A star shines by utilizing the energy of the nuclear reactions going on within it. I shall consider three important cases of what can happen when this source of energy runs out. (i) If the star's mass is less than about 1.4 solar masses it shrinks to become a *dwarf*, the gravitational energy generated by the collapse being converted to heat. So it still radiates more or less white radiation similar to that of the sun: it is a *white dwarf*! Later it is believed to shed its outer layer and its light becomes very faint. In due course it cools to become a *black dwarf*. (ii) If the star has a mass exceeding about five solar masses, the collapse resulting from the exhaustion of the nuclear fuel triggers off a terrific explosion which again leads to the shedding of the outer layer. The gravitational forces in this highly compressed state are enormous, so that not even light can escape from the remnant, which becomes a *black hole* (p 194). (iii) For stars of intermediate mass, electrons and protons are pressed together to combine and so produce neutrons and neutrinos. The latter do not interact strongly and escape, while the neutrons form a new nucleus for a *neutron star*. Eventually the pressure on the neutrons (which do not like being close together) is so great that the star blows itself apart. During an explosion which again expels the outer layers, and is of a brightness equivalent to many million suns, a *supernova* is generated. The densities of matter which occur in these compressed objects are inconceivably high: typically one thousand million metric tons in a volume of one cubic centimetre!

These large astronomical objects are of great interest. Thus in 1054 AD Chinese observers noted a bright star, now known to have been exploding. It has spread out in the meantime and has become a more diffuse nebula which is still the source not only of light, but also of x-rays and of radio waves. It is a 'supernova remnant', called the *Crab* nebula. Many supernovae are now known. They are very bright and have been classified by cosmologists. Because of their brightness they can be very useful in connection with the hard problem of determining *astronomical distances*. This requires objects of roughly constant intrinsic brightness so that their apparent brightness can serve as an indicator of distance. These supernovae are referred to as *standard candles*. Teams of astronomers work on them, e.g. the High-Z Supernova team of Cambridge, Massachusetts, or the Supernova Cosmology Project of Berkeley, California. They have found of the order of fifty.

Since Hubble's time *quasars* have been discovered. Their light is strongly red shifted and they are believed by some to be the most luminous as well as the most distant objects which we can see. Some astronomers, however, doubt that quasars are as distant as suggested by their red shift. More than 5000 quasars are now catalogued and their red shifts are large, ranging from about $z = 0.1$ to about $z = 4.9$. Astronomers have made the *'cosmological hypothesis'* that these values are due to the Hubble expansion. Hubble's law, analogous to figure 7.2, holds for them, but with more scatter than usual. Some objects with $z = 10$ have recently been seen. We are looking into the past! This light must have been emitted when the universe was only 9% its present size: there was then no solar system, no earth of course, which all developed while these photons were on their way to the spot where the earth would eventually be.

The Hubble *parameter* (it cannot be regarded as a *constant* in time) is usually denoted by H. It relates speed to distance, such that $1/H$ is actually an estimate for the age of the universe (see p 170 and figure 7.4, p 181). If the faster objects are always further away from us, one may suppose that the reason is that they all started in the same general region. Therefore the Hubble expansion suggests a *Big Bang* which started it all off, a matter to which I shall turn in section 7.3.

Astronomical measurements have shown that of 500 nearby stars many have orbiting companions roughly the mass of Jupiter. Among the smaller objects of interest are *comets*. They do not satisfy Hubble's law, can be found just about anywhere at any time, and they give the amateur astronomer a chance to make his or her name. Thus Alan Hale (New Mexico) and Thomas Bopp (Arizona) found, in July 1995, the highly visible Hale–Bopp comet. This is estimated to give off thousands of kilograms of matter every second. Its nucleus alone has a diameter of the order of 40 kilometres with extensive surrounding matter. One significance of comets resides in the marginal possibility that the oceans, and possibly primitive life, were started by meteorites and comets hitting the earth in large numbers. This idea, which has important backers, is, however, not generally accepted. Meteorites are commercial business—it is one way to have a bit of the moon (say) in your home. Michelle Knapp's car in New York State was reputedly hit by a 12 kg meteorite and, with meteorite included, its value soared to $69 000!

Still smaller objects include *spacecraft*. Pioneer 10, for example, launched in March 1972, is now at a distance of ten thousand million kilometres, or at a distance equal to 67 times the mean earth–sun distance. Twenty-six years after its launch its power due to radioactive pellets has become too weak to make it worthwhile for the American space administration to maintain contact.

There are three points which I ought to emphasize. First, the general expansion does not imply the expansion of atoms, stars and galaxies themselves. If they change in size, this is due to entirely different mechanisms such as *gravitational contraction*, evaporation, etc. Secondly, the universe does not expand into empty space. The expansion *creates* its own space. This is a concept from general relativistic cosmology and lies outside Newtonian physics. Lastly, we see that in cosmology it is not all plain sailing as far as scientific agreement is concerned, as already hinted at in the title of the book cited under [7.3]. We have also seen this in our discussion of distances and of the ages of celestial objects.

7.3   Cosmological models

Modern scientific cosmology started in 1917 when Einstein applied the general theory of relativity to the cosmos. He found to his dismay that if he treated the matter content as uniformly smeared out the system would not be stable. This was in contradiction with the belief of the time. The universe was regarded as static by everybody. This was very understandable since Aristotle had already noted that our image of the stars does not seem to alter in any crucial manner as time goes on.

Thus Einstein (box 7.2) had a choice: either accept his equations and propose that the universe is not a static system, or hypothesize a force of universal repulsion which could oppose the universal attraction of gravitation and so yield a static universe. Both steps would have been at the time highly revolutionary. He favoured the static universe, and thus was born the *cosmological constant*. It is possibly related to a fifth force (see p 55), called quintessence, which is perhaps related to the energy density of the vacuum. Had he taken the alternative view he could have predicted the expanding universe 12 years before Hubble inferred it from astronomical data. But he did not. That is why he regarded his decision later as his 'biggest mistake'.

The Russian scientist A A Friedmann (1888–1925) showed in the years 1922–24 that the Einstein cosmological equations could be solved to show that the uniformly smeared out matter in his model universe had to expand or contract. It could not be static. We see that in this instance theory was actually ahead of experiment and a new subject was created: we now had to investigate the changes in the universe. But do not believe that the cosmological constant was now dropped. Far from it, it gave theoreticians another parameter to play with, and it had its ups and downs in the following decades. Possible models of the universe were investigated and classified often with the cosmological constant included. Positive cosmological constant implies the existence of the universal repulsion envisaged by Einstein; it endows even empty space with an energy density and aids the expansion of the universe. This tends to 'unbend' the scale factor curves of figure 7.4, thus increasing the inferred age of the universe. It has been found to be a useful hypothesis which ensures that stars are not found to be older than the universe! Recent supernova studies suggest that there is indeed some effect at work which counteracts gravity, and which might be the result of a positive cosmological constant. Occasionally the cosmological constant is taken as negative, which encourages the occurrence of an oscillating model (curve 3 in figure 7.4).

Box 7.2 Einstein (1879–1955; NL 1921).

In 1905 the special theory of relativity gave to the marriage of energy and mass (see p 21) a more spiritual—should we say platonic?—and more permanent form. This had two effects which are of interest here. The first was to recognize mass as yet another form of energy, using the speed of light (usually denoted by 'c') to convert mass to energy. This was largely due to Einstein. The Nobel prize, though not given for relativity theory, was a considerable advance on being an Examiner for the Swiss Patent Office in Bern, making ends meet by giving, in addition, private lessons to students of Mathematics and Physics, and to make this more attractive by offering free trial lessons (figure 7.3)! Einstein is the *hero* of this chapter.

Although equations are more or less banned from this book, poems are not. So I offer, in elucidation of the last paragraph, part of a poem by Sagittarius of the *New Statesman*, 27 July 1947:

> *The final truth is stated*
> *With certitude emphatic,*
> *All doubt is dissipated*
> *In symbols mathematic.*
> *Man reaches his ambition,*
> *The confines of cognition,*
> *The cosmic secret's bared,*
> *And here's the proposition—*
> $E = mc^2$.

You may say, to my remarks about interference etc, quite rightly, that some old problems seem to arise again and again! In the 17th century one wondered whether to treat light as particles or waves, and in the 20th century the same query arises in a somewhat different form, for, as we saw in section 3.8, photons are 'wavicles' (to use this old-fashioned term, of p 46), since they have both wave and particle properties. But we have agreed to call these entities simply 'particles'. We see once again that our beautiful and well-defined concepts are not ideally adapted to describe this actual and complicated world. We may take comfort from the thought that even Einstein claimed as late as 1951 in a letter to Michele Besso 'All these fifty years of conscious brooding have brought me no nearer to the answer to the question 'what are light quanta' [7.5]. A few physicists actually want to use the terms 'light' or 'radiation' instead of 'photon', as being less misleading [7.6]. When Einstein gave the photon interpretation of the photoelectric effect (see p 160), one could hear Newton laughing in his grave 'I told you so—it *is* a particle'. Einstein was puzzling about the quantum theory until his last days. I met him (and his assistant Helen Dukas) in 1953/54 in his house in Mercer Street in Princeton. A surprisingly tall man, he entered the room in his dressing gown, ready to talk, but only about quantum problems.

Figure 7.3 An advertisement inserted by Einstein in the '*Anzeiger fur die Stadt Bern*', dated 5 February 1902. The Albert Einstein Gesellschaft and Dr A Meichle, Bern, are gratefully acknowledged for permission to use this figure.

Just as we had a taxonomy of particles in section 3.9, so we can have a taxonomy of universe models based on Einstein's equations. They are called *Friedmann models*. Do not let us get carried away and talk about 'the universe'. I am afraid all we can hope for is to devise good 'models' of the universe. It was Georges Lemaitre, the Belgian Jesuit priest, who made the connection between the Friedmann models and the Hubble expansion.

In these models the size of the universe is fixed by a *scale factor R(t)* which depends on the time. It scales up all distances between astronomical objects appropriately. Each model has its own scale factor curve showing how its size changes with time. There are then two main possibilities for the model: indefinite expansion or recontraction. In all these cases the Hubble law holds: the velocity of recession (or of approach in the case of recontraction) is *at a given time* proportional to the distance from a given galaxy. But the Hubble parameter itself will in general change with time. The straight line corresponds to a model in which there is no slowing down in the expansion (figure 7.4). Such a model contains a negligible amount of gravitationally attractive matter. This *Milne model* is named after the early Oxford cosmologist Edward Milne (1896–1950) who, incidentally, did not accept the general theory of relativity, but devised his own theory (which is not now accepted). All these models start with $R = 0$ and are therefore 'Big Bang' models.

The bend of the curves shows that the matter in the model universe slows down the expansion by virtue of its gravitational attraction. Thus if the model universe is now at the point P its greatest age would result from the assumption of negligible gravitational attraction. The increment c of the figure has actually the value $1/H$, mentioned on p 175. Curve 2 is similar in shape to what you would find on the assumption of a very popular model associated with the names of Einstein and de Sitter. As you pass from curves 1 to 2 to 3, the gravitational slowing of the expansion becomes more and more important. The border line between recontraction and indefinite expansion occurs at a *critical* average mass density, close to the present value of the Hubble parameter, and comes to about five hydrogen atoms in a cubic metre. This is the mass density for the so-called *Einstein–de Sitter* model, whose Hubble parameter decreases to zero as time marches on, and which only just manages to expand for ever.

Here is an easy problem for the reader: could there be curves in figure 7.4 which bend upwards, suggesting that the speed of expansion increases? (Answer: see end of chapter).

The Hubble plots for supernovae contain astronomical observations blended with some theory, so that they do not represent purely empirical data. But in any case it is interesting that they turn out to be reasonably consistent with the Einstein–de Sitter, and even the (almost empty!) Milne, model. The interpretation of these plots is of course influenced by the presumed value of the cosmological constant.

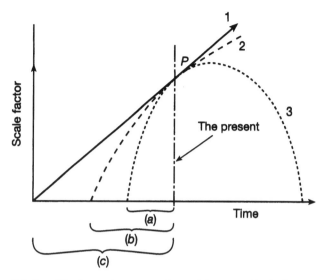

Figure 7.4 Possible dependences on time of the cosmological scale factor. (*a*) denotes the age of the universe if there is a recontraction. (*b*) denotes the age of the universe if there is indefinite expansion. (*c*) denotes the age of the universe in the absence of deceleration.

There are three main parameters to describe Friedmann models quantitatively: the Hubble parameter, the average matter density and a quantity giving the deceleration due to gravitational attraction of the expansion. They all depend on time, so that *if* we talk about their numerical values, we usually mean their *present* values. A fourth parameter is also used: namely the cosmological constant, which we met at the beginning of this subsection. Perhaps it is not a constant but depends on time! If it had been very large at a very early stage, as has also been envisaged, it would have caused a very rapid early expansion which would have smoothed out the initial chaos of the Big Bang and given us the rather smooth universe which we see today. An early rapid expansion of this type is in fact what is proposed in the so-called *inflationary model*. Some cosmologists believe that the expansion of the universe is still accelerating, and this may be due to a sufficiently positive cosmological 'constant'.

I shall now describe another model. At several periods in history it has been thought that some object was uniquely qualified to be considered the centre of the universe: the earth, the sun, and our own galaxy have been in that position. But they were not unique: there are several planets, the sun is one of many stars, and there are many galaxies. So cosmologists, up to say 1950, took it as an axiom that

> The large scale appearance of the universe is the same at a given time for all observers located in, and moving with, galaxies.

This is the *cosmological principle*. It asserts that on the largest accessible scale the universe is homogeneous. On the smaller scale of galaxies or clusters of galaxies it is of course inhomogeneous. That it looks roughly the same on many different scales is an additional thought and may be expressed by saying that it is *approximately* 'fractal' or 'self-similar'. On the word 'approximate' hangs a controversy [7.7]. The reader should look for beautiful fractal pictures in the literature.

Further, if an observer is *not* moving with a galaxy he or she could see a violation of the Hubble law which other observers accept. But in fact it has been shown that if the universe has broadly the same properties in all directions and obeys the cosmological principle at all times, then Hubble's law is valid at all times (see, e.g., [7.8]). You can see this by marking crosses on a balloon to represent galaxies, and then blowing it up. The crosses will all separate from each other. The universe is subject to a Hubble-type expansion, as viewed from any of these galaxies. Of course this has to be translated from the surface of a balloon to three dimensions.

In recent decades increasing experimental evidence has often made it unnecessary to appeal to the cosmological principle, rendering it a little out-of-date. Note that it tells us something along the lines of the much older Copernican principle (p 226), which says that we are in a typical location in the universe, and not in a specially favoured position, as was advocated by Ptolemy, the Greco-Egyptian astronomer who lived about AD 140.

A very simple cosmological model is obtained by extending the cosmological principle to be valid *at all times*. It is then called the *perfect cosmological principle*. We look around ourselves in the model smeared out universe and since everything looks the same, we do not know where we are. This is broadly speaking correct. But if the perfect cosmological principle is adopted, we do not even know what the time is! The resulting theory is called the steady-state model of H Bondi, T Gold and F Hoyle.

In this model the density of matter must remain the same as expansion proceeds, so that you have to have universal matter creation. This turns out to be so small (in a century one atom has to be created in a cubical box whose side is one kilometre!) that it cannot be detected by present methods. Still, on the basis of this model the creation of matter is not confined to the remote past, but continues in the present, thus violating energy conservation. The age of the universe in the simplest form of the model is indefinite, so that it does away with any poorly understood initial explosion. Further, the Hubble parameter is in this model independent of time: it really is a constant! Attractive though this model is in its conceptual simplicity, experimental results no longer favour it in its original form. Amendments are available, but do not enjoy wide support.

Box 7.3 Cosmological controversy.

There was quite a discussion about these models in which notable cosmologists participated. Herbert Dingle (1890–1978) was particularly active. Although President of the Royal Astronomical Society from 1951 to 1953, he questioned the validity of the special theory of relativity in his later years. His arguments with famous cosmologists like Fred Hoyle and William McCrea gave rise to the following verse [7.9]

The ears of a Hoyle may tingle,
The blood of a Dingle may boil,
When Hoyle pours hot oil upon Dingle,
And Dingle cold water on Hoyle.

But the dust of the wrangle will settle—
Old stars will look down on new soil—
The pot will lie down with the kettle,
And Dingle will mingle with Hoyle.

7.4   The 'relic' radiation

I have noted that the universe is roughly uniform with similar properties in all directions. This so-called isotropy is associated with an important and intriguing story. In 1964 a rather unusual radio antenna at the Bell Telephone Laboratory was to be used to measure the intensity of radio waves emitted by our own galaxy. This difficult measurement required high accuracy and hence the elimination of any extraneous disturbance ('noise'), which always has a component due to the thermal motions of atoms and electrons, which, as we saw on p 12 and p 154, can never be at rest. This disturbance is measured in terms of the noise which would be produced in a box whose walls have a certain temperature, and which therefore contains black-body radiation at that temperature (p 149). The antenna noise was found to be equivalent to about 3.5 K, i.e. small but persistent. It could not be eliminated even after the repeated removal of some pigeons which had nested in the antenna roof!

Now switch back to cosmology. If Einstein was the father of relativistic cosmology, the American George Gamow was his main disciple. He had suggested in the 1940s that the radiation left over from the Big Bang would spread out with the universe, and in spreading out it would cool, rather like the steam coming out of the spout of a kettle. The laws for this were well known, and at the present time this radiation should be at about 5 K. It is just what the Bell antenna had detected. This interpretation required (among other things!) good knowledge of the literature! In the ensuing discussions Robert Dicke (1916–1997) and James Peebles, both of Princeton, played a prominent part. The most widely used name for this radiation is: *cosmic microwave background* (CMB).

The cosmological interpretation was accepted and the authors of this 1965 experimental paper, Arno Penzias (b 1923) and Robert Woodrow Wilson (b 1936) received the 1978 Nobel prize. One reason was its importance in distinguishing the Big Bang theory from the steady-state (or continuous creation) model. The latter was in strong competition with the former because of its elegance, but it did not contain a prediction of a residual (or 'relic') background radiation in its armoury. So it now became less popular among scientists [7.10]. The energy in the various frequency intervals of the 'relic' radiation has been checked and was found to have all the characteristics of black-body radiation. It was so accurate that when John Mather presented these observations to a meeting of the American Astronomical Society in January 1990, the audience gave him a spontaneous ovation.

We now have as an entirely new feature of the universe the notion that all of it is bathed in this radiation at a temperature of 2.7 K. The *critical density* of the Friedmann models corresponds to about 5 hydrogen atoms per cubic metre, and the energy density, in terms of equivalent mass, in the background radiation is less than that by a factor of about 2000. Astrophysicists can now have a great deal of fun, as the thermodynamic properties of black-body radiation have been well known since the beginning of the century, and could thus be readily calculated, subject to allowable simplifying assumptions. For example, its entropy is readily worked out. With assumptions of how to treat the smeared out matter component of a homogeneous model universe as well, you could estimate the thermodynamic properties of these models, including not only their energy content, but also their entropy.

If the cosmic microwave background originated with the Big Bang, it would be expected to be pretty isotropic by now. Indeed if its temperature is measured in various directions, you arrive at the same temperature to an accuracy of one in one hundred thousand. But one expects a variation in the spectrum, when taken in different directions, due to the motion of the earth in this cloud of radiation. This is experimentally estimated to be 600 kilometres per second (one million miles per hour), confirming theoretical results.

It is very likely that some of the matter in the universe does not emit light. Why should it? You would expect interstellar gas, gravitational waves which have, of course, also got an equivalent mass density, neutrinos (which appear to have some small rest mass), etc, to add to the effective mass in the universe. This is common sense, but as we know only too well, common sense will not do. Any proposal for such *dark matter* has to be integrated most carefully into the cosmological picture as a whole. Is the interstellar gas hot? Then it will in fact radiate. Is it very dilute? Then its radiation may not be visible. Actually, the existence of dark matter has been inferred from the motion of galaxies in clusters of galaxies by gravitational effects. If dark matter exists, then it tends to bend the curves of figure 7.4, thus shortening the *inferred* age of the model universe. The age of the universe, as obtained by other methods, acts as another control. The current balance of opinion is that the density of matter is of the order of the critical density, but that roughly 90% of it is invisible. This notion is based on both theoretical predictions from particle physics and from studies of the gravitational attraction in our universe. The investigation of this problem has spawned experiments by the UK Dark Matter Consortium. This places detectors in shielded underground places to search for these additional cold, weakly interacting particles (*wimps*).

The general theory of relativity also predicted the existence of *gravitational waves*. According to this concept, each time you raise your arm gravitational waves are emitted. However, the energy involved is tiny. These waves have been identified indirectly—their direct experimental observation is still too difficult. The identification has been achieved by virtue of regular pulses of radiation sent out by a pair of compact stars consisting largely of neutrons and rotating about each other. They belong to a class of stars named *binary pulsars*. The pulses from the pulsar under observation matched very well what you would expect from the change in the received radiation due to the loss of energy by the pulsar by virtue of the emission of gravitational radiation. This basic work was done by Russell Hulse (b 1950; NL 1993) and Joe Taylor (b 1941; NL 1993).

The first pulsar was discovered in 1968 by Anthony Hewish and Jocelyn Bell and the regularity of the pulses suggested the possibility of signals from 'little green men' (LGM), i.e. from intelligences in outer space. This often told story was soon dismissed and we now interpret *pulsars* as rotating neutron stars. More than 900 pulsars have been found in the meantime. They are sometimes referred to as celestial lighthouses: we receive the regular flashes of radiation as the beam from the neutron star sweeps again and again across the surface of the earth.

Gravitational waves will be most pronounced in those parts of the universe where vast concentrations of matter heave around. Unfortunately these absorb the radiation due to light, radio and x-rays, which have been used so far to study our surroundings. They have yielded an optical universe, an x-ray universe and a radio universe. If we could use gravitational waves to study it, we might one day add a gravitational universe.

Among the rich rewards of the study of the microwave background, the departures from isotropy, i.e. its anisotropies, should be mentioned. Its slight temperature variations across the sky have been analysed by delicate theoretical and experimental methods and have led to constraints on possible values of cosmological parameters. These fluctuations (*ripples* in cosmic background radiation) tend to clump gravitationally, then they *pull in* interstellar gas and form early forms of galaxies ('protogalaxies'), and eventually these interact to form the larger structures we see today. This earlier evidence can be seen only if we look far enough back into the history of the universe, i.e. at more distant structures. The procedure of gaining an insight into the history of the universe is rather like reconstructing the history of a town from very early paintings and maps, taking account of human activities. In the heavens we have instead the whirlings of clouds of matter and radiation at such a great distance from us that they correspond to a much younger universe, but all dominated, then as now, by gravitational attraction [7.11].

7.5 Olbers' Paradox

In 1993 a former President of the Royal Astronomical Society pointed out that from the mass of astronomical observation it could be that only 'two and a half facts' are really vital for cosmology [7.12]. One of these, I am glad to say, I have covered already. Fact 1: the galaxies are receding from each other as in a uniform expansion. Fact two and a half is fairly commonsensical: the contents of the universe have probably changed as the universe grew older. The remaining fact 2 is: the sky is dark at night. While obvious, it has an interesting history and its explanation is, surprisingly, somewhat controversial.

Are these really facts, though? From what we have seen, it could conceivably be argued in the future that the red shift is not due to recession. In saying this, I expose myself to the charge of logic-chopping from the cosmology establishment. All findings are subject to correction they will rightly say, as I have indeed emphasized throughout. However, I must be honest. Should, in the year 2005, the newspapers carry headlines saying that the cause of the red shift is not a recession of the galaxies, but is gravitational (say), and the universe is not expanding after all, I do not wish my readers to write to me postcards telling me that I have misled them. It would make me very unhappy. Actually this is not very likely, because of the interlinking arguments of the cosmologists. However, I wish to draw attention to this far-out possibility.

Although the paradox of the dark night sky can also be attributed to P L de Cheseaux who noted it in 1744, it carries the name of the German astronomer H Olbers who discussed it in 1823. The invention of the telescope showed that there are many stars and it was assumed that they are fairly uniformly distributed. This has now been shown to be valid with great accuracy. But it was realized that if you place a *finite* number of stars into a static model universe, they will attract each other gravitationally and ultimately make a big lump. A more attractive picture for the 19th century was therefore to adopt the model of an *infinite*, static and uniform universe. This could be stable, but leads to another paradox! If the stars have been there all the time, then a line of sight, drawn in any direction from an observer, must end on the surface of a star so that the sky should appear in all directions with the brightness of a typical star, like our sun. This is not observed!

There are several ways of avoiding Olbers' paradox. The argument itself is actually wrong: a static universe, as imagined at the time, does not contain enough energy to produce a bright night sky [7.13]. This point has been ignored by many commentators, who have given the red shift due to expansion as a resolution of the puzzle. It was argued that the recessional weakening of photon energies would reduce the night sky to darkness. Of course it would be attractive to argue that Olbers might have predicted the expansion of the universe via the Doppler red shift, already known at his time, in order to avoid his paradox. Romantic?—certainly: Olbers steps outside at night and says: 'It is dark—by Jove, the universe must be expanding!' But unfortunately untrue. A better explanation is that stars have a finite life and shine for only a limited period. That is why the Olbers argument fails. The nuclear fuel which powers the heat and light of our sun and of other stars, eventually runs out. (It may take ten thousand million years before this happens in the case of our sun.) Some stars have started to shine only recently and their radiation has not yet spread out sufficiently. The universe (in that sense) is too young for Olbers' paradox. For its history see, for example, [7.14].

7.6 The oscillating universe

In an oscillating universe model the entropy can return to its original value, as if to verify some Pythagorean or Mayan philosophy of cyclical time. Abandoned to distant dreams, let me briefly consider life in a contracting universe. It is of course possible that there is no contracting phase but indefinite expansion, or that there can be no life in a contracting phase. This is a first possibility.

But a more interesting second possibility is that fairly normal life is possible in a contracting phase. Humans would regard the direction of time towards smaller entropy states as 'the past', and that towards higher entropy states as 'the future'. On that basis, if the entropy in the contracting phase returns to the value it had at the beginning of the cycle, we would have the situation illustrated in figure 7.5, with the amusing consequence that human beings would still regard the universe as expanding in the contracting phase. This would suggest the rule 'Living things can *never* see a contracting universe' [7.15]. It is then not possible to determine whether an oscillating universe is 'actually' in a contracting or expanding phase. This decision requires our demon number 6 (p 78). This situation is of course trickier if our simplifying assumptions are removed. (Hawking thinks he has made a mistake in this connection [7.16].)

Consider, therefore, a more realistic model, namely a universe consisting of matter and radiation. Each of these two phases is in equilibrium by itself, but not in equilibrium with the other one. Heat is then transferred from the hotter to the colder, and as a result the entropy always increases, and so does the energy. This is possible because energy conservation in the usual sense does not hold in general relativity. Suppose next that at the end of the cycle another one can start by cutting out in our theory a short period of maximum compression. Let us start the next cycle with the same energy as was found at the end of the previous one and with equal and opposite speed. We thus start it with an expansion. We can repeat this process and then we have a rough model of an oscillating universe which can go through many cycles [7.17]. The calculation starts at a high energy density at the last Big Bang and shows that the density of the universe decreases towards the *critical* density (p 180) after a large number of cycles: typically to an accuracy of 1 in 10 000 after 1000 cycles. The interacting phases require more work for the recontraction, thus lowering the energy density. The quantity governing the extent of the universe, the *scale factor*, might behave as shown in figure 7.6.

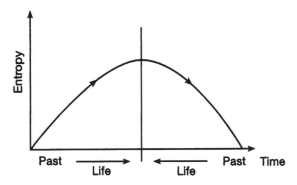

Figure 7.5 A contracting phase follows expansion. The arrows below the entropy curve give the direction of human time. The arrow on the curve gives the time as it would appear to a disembodied intelligence.

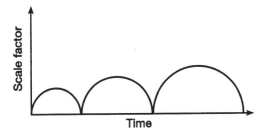

Figure 7.6 The time dependence of the extent (i.e. of the scale factor) in an oscillating universe.

The above picture is pretty rough, particularly since it does not give us a clue as to how the model can proceed from the *Big Crunch* at the end of the contraction to the next *Big Bang*. Or, to use cosmologist's jargon, how the model can *bounce*, and can do so repeatedly. Nevertheless it gives a hint of a possible reason why the energy density of the universe is very close indeed to the critical density. This result is found from studies of the early universe, incorporating the grand unified theory noted in section 3.10. The question of why this should be so has been named the *flatness problem*. The above reasoning suggests one possible (but not universally accepted) answer: 'Because the universe has already gone through many cycles'.

There are more popular proposals. In one of these, the *inflationary model* of p 181, the universe suffered a very rapid expansion. The factor of size multiplication involved is many million times what is considered in the more standard Big Bang model. It is presumed to occur at even earlier times than are incorporated in the outline history of the universe given in figure 7.7, namely about one million million million million millionth of a second after the Big Bang. The cause is believed to be that gravity was in effect repulsive for a brief period. Each of several variants of this model furnishes a popular way of solving the flatness problem mentioned above. However, difficulties remain. Some of these are related to the precise cause of this sudden and explosive expansion. Others are due to the fact that the statistics of the cold and hot spots, i.e. of the density disturbances in different parts of the sky, are not always as random as suggested by the inflationary model.

In figure 7.7 radiation and matter in relativistic motion are grouped together because they have rather similar properties and it is therefore hard to distinguish one from the other. This figure shows how the universe, originally dominated by this radiation, now contains mainly matter in non-relativistic motion.

In the process of bouncing, the laws of physics might of course get changed completely, so that we emerge with a different physics. Apart from noting this scenario, there is nothing more we can *do* about *this* possibility.

Other problems plaguing contemporary cosmology include the *horizon problem* that the 'relic' radiation is very homogeneous on a much more extensive scale than we would expect. The difficulty is that the points whose radiation we receive must have been in communication in the past to create the impression of homogeneity, and it is difficult to see how this can be for points that may have been arbitrarily far apart. Various ways of overcoming such problems have been suggested and their popularity waxes and wanes depending on astronomical observations. But an inflationary-type model is currently most popular.

The cosmological and thermodynamic arrows of time seem to be superficially contradictory, since we would expect a high degree of 'disorder' near the beginning of the Big Bang, while 'order' is expected to emerge later with the formation of galaxies, etc. The cosmological process, like the biological one, *seems* to be anti-thermodynamic, as already noted in section 5.6. Two observations help with this difficulty:

(i) Ordinary finite system which are driven far from equilibrium display surprising reservoirs of structure which are not expected if attention is confined to equilibrium. This applies, as was seen in section 5.4, to fluids and chemical reactions as well as to semiconductors. This appears to be the main way around that difficulty.

(ii) Gravitational effects keep the universe and many of its component parts away from equilibrium. Ordinary thermodynamics therefore needs amendment before it can be used. Actually, gravitational thermodynamics, though it has been studied for a long time, is not yet a well-developed discipline. Already the beautifully simple-sounding system of a self-gravitating sphere of black-body radiation leads to quite involved results, for example for its heat capacity as shown by Sorkin, Wald and Jiu in 1981.

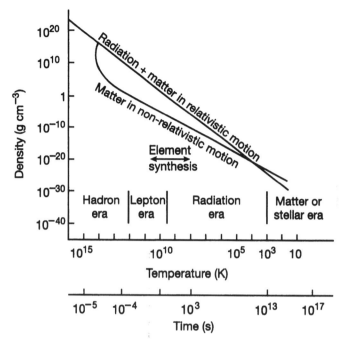

Figure 7.7 A rough outline history of the universe in term of its temperature, time and its energy plus matter density.

7.7 The origin of the elements

The Big Bang theory of the development of the universe has three main items of support: the expansion of the universe, the relic radiation and, to be discussed next, the *abundance* of the elements. The abundance of a chemical element is the fraction of the mass of the element present in a system as a fraction of the total mass present. Today's abundances in the universe have been estimated from various analyses: from the composition of rocks on earth, from stellar structure, from cosmic rays and generally from the spectra of light received on earth from various sources.

A theoretical estimate of these abundances can be obtained from a model of an early universe which is so hot that any complete atom is not only stripped of its electrons, but even its nucleus has broken up. The mass equivalent of the radiation energy in a cubic millimetre could be as high as 10 million kilograms. One millionth of a second after the Big Bang, the temperature is still believed to be of the order of a thousand million million degrees and the contents of the universe is basically a soup of photons, quarks, neutrinos and their antiparticles (figure 7.7). Then electrons appear and protons and neutrons are later made from quarks. Three minutes after the Big Bang the lighter elements of table 6.2 and their isotopes appear. The heavier elements come later, but only in relatively tiny amounts. The heavier elements are believed to be produced in the stars, in *stellar* as against *cosmological nucleosynthesis*: the stars are the furnaces for their production. For example, if a supernova blows up (see p 174), interstellar space receives some of these heavier elements.

In order to obtain abundances, you have to adopt a theoretical model for the Big Bang by assuming an initial radiation density and an initial temperature. After that one feeds into a computer the important nuclear reactions which occur at various temperatures and densities. The gradual build-up of the elements through more than hundred nuclear reactions can then be traced. The resulting model of the cosmological nucleosynthesis gives something like figure 7.8. It shows that there exists a density of the universe, indicated by an arrow, for which the abundances are in reasonable agreement with the observed ones. Helium, at about 23%, is our most abundant element after hydrogen at 76%. Even about 27% of the mass of the sun is helium. The spectra of comets have also been studied and show that they are made of the same stuff as the earth in the sense that the isotopic ratios are much the same.

7.8   Black holes

There are two rather elementary 'laboratory experiments' for cosmology. (i) You blow up a balloon with crosses marked on it to illustrate that Hubble's law can hold for each galaxy separately (see p 182). (ii) You throw a piece of chalk up and catch it on its return. This is analogous to a recollapsing universe: the attraction from the earth is too effective to let the piece of chalk reach outer space. How fast must you throw this piece of chalk, so that it can in fact leave the attraction of the earth? This speed is called the *escape*

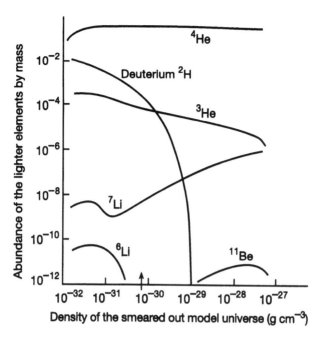

Figure 7.8 The abundance by mass of the lighter elements as a function of what value we assume for the present smeared out density of the model universe.

velocity and for the earth it is about 11 kilometres per second. If you throw the chalk a little harder so that it goes through the roof, and indeed never comes back, you have reached or exceeded this velocity. This is analogous to an indefinitely expanding universe: the universe is then not massive enough for gravitation to stop the expansion. It has been suggested that the universe is expanding at about twice its escape velocity.

Let me take the velocity of sound in dry air at normal temperature and pressure as a unit, called the Mach, after Ernst Mach (1838–1916). The escape velocity for the earth is then 33.7 Mach, i.e. much faster than current jet planes like Concorde. The bigger the object, the greater the escape velocity. For the sun it is about 1861 Mach (618 kilometres per second).

For objects of sufficiently large mass the escape velocity reaches or even exceeds the speed of light. In that case no piece of chalk, however fast it is thrown, nor even light, can escape. Such objects can therefore not be seen optically and are called *black holes*. They can be identified in the main by their gravitational attraction. Thus it is very hard to provide observational proof that a black hole has been found. One's talk is restricted to black hole 'candidates' (BHC). Many astrophysicists want to discover the best contender. Notable ones are certain X-ray sources (LMCX-1 and LMCX-3) in the Large Magellanic Cloud as well as the binary X-ray source CygX-1 in the Swan constellation, discovered in a 1965 rocket flight, and situated in our galaxy. We may add 'the least unreliable' candidate, named A 0620-00 [7.18]. Note a recent table of ten candidates [7.19].

The jiggling of particle–antiparticle pairs brought about by the extremely strong gravitational field at the surface of a black hole leads to the emission of radiation from the *surface* region. This is called Hawking radiation, after its discoverer, and it leads to a gradual loss of mass of the black hole. The radiation emitted has qualities which enable it to be regarded like the radiation from a black body, so that we can associate an absolute temperature with the simpler forms of black holes. In fact these black holes are simple because their properties depend essentially only on three variables: mass, charge and angular momentum. There are various other types of black holes which are more complicated and have been discovered by theory alone.

Experimentally we would expect black holes to be formed from inhomogeneities at a time just after the Big Bang. These are called primordial and may have largely disappeared by evaporation. Then there are black holes which result from stars at the end of their lives. Lastly supermassive black holes are suspected at the centre of some galaxies. Their masses would be of the order of a million solar masses. There is still some element of doubt about their existence. For this and many other topics covered in this chapter the reader is referred to some recent fine expositions, for example [7.20].

Normally we think of a body which emits radiation or heat as cooling down as a result. We say that it has a positive heat capacity. Once gravitation is involved, however, some more surprises are in store for us. A star, powered by its nuclear reactions, emits light and heat. At the same time it contracts under the influence of gravitation. The stellar material thus gets hotter. The result? It becomes hotter on emitting energy!! Thus one has a *negative* heat capacity in this case.

Black holes can exhibit the same surprising phenomenon. The simple black hole we have been discussing has a temperature which increases as it loses mass: it becomes hotter and hotter. The hotter it is, the more energy it emits. So it stands to reason that it must evaporate and end its life with a kind of explosion. Conversely, the more massive a black hole the slower it evaporates. It also has a negative heat capacity. From a black hole and its temperature it is possible to infer an entropy. Like all thermodynamic quantities it should be calculable from the principles of statistical mechanics. This has presented many problems, which seem to be currently in the process of being overcome.

Black holes can in theory also rotate. Energy can be extracted from such a system by firing a block of matter at it in an appropriate manner, i.e. with a suitable speed and in a suitable direction. Then things can be arranged so that part of the block enters the black hole, while the remainder emerges with *more* energy than the whole block had originally. The gain of energy is at the expense of some mass and some angular momentum of the rotating black hole. Here is a method (the *Penrose mechanism*) of extracting energy from a black hole system. There has not been an occasion to try it out as yet!

7.9 Some problems

It must be clear by now that in the great issues of cosmology there are and will always remain uncertainties, as well as topics which are essentially matters of opinion. Not even the steady-state model is fully discarded. This is so in spite of Martin Ryle's observation of 1955 that the early universe had a higher density of radio sources than we have now, so that the steady-state theory did not seem tenable. Then there is the idea that photons lose energy as they travel through space, so that the cosmological red shift might not be due to a recession effect. This idea is frowned upon for various astrophysical reasons. But could this red shift be due to the gravitational effect already explained on p 159? Probably not. But it is unlikely that any of these matters can be decided with certainty, and the science of cosmology can be expected to make pronouncements on these issues only in the form of probability statements.

In fact, a well-known cosmologist has ventured beyond what I am willing to do here by giving actual probability estimates as follows: chance of correctness of the Big Bang model for the period of one second to the present: 99%. Chance of correctness of the idea that very soon afterwards the universe went through a very rapid expansion, known as inflation: 30%. Chance for the *string* or *superstring theory* (see p 48) 10%. Chance of success for certain other ideas not discussed here, namely Penrose's *twistor cosmology* or Hawking's *no boundary* proposal: 1% [7.21]. The latter is based on the following simple idea: if we were two-dimensional beings, confined to live on the surface of a balloon, then we would know no spatial boundaries to our 'universe'. Generalize this first to three-dimensional space and then, in a further step of imagination, also to time. The accompanying mathematics is, of course, complicated.

It is hardly surprising, then, that we can put a finger on a number of questions to which answers are not yet known. We have noted, p 191, the flatness and the horizon problems. Here are some additional ones. But the reader must always keep a clear distinction in his/her mind between the real universe and the model universes created by theorists.

(i)   The neutrino has a non-zero rest mass (p 59). This would account for a great deal of the 'dark matter' for which we have evidence when fitting together cosmological data.

(ii) Why is there practically no antimatter in the observed universe?

(iii) Is the best model universe one that is always expanding, or does a re-contraction furnish a better model? Or is the averaged matter density actually 'close' to the critical density?

(iv) How are galaxies formed from initial fluctuations in the 'relic' radiation?

(v)   What of the cosmological constant: is it small or actually zero?

(vi) Can we give some explanation of the constants of nature, including particle masses?

(vii) Could our universe be part of a giant fluctuation in space and time?

(viii) Might there have been several 'mini-bangs' ?

(ix) What is the lifetime of a proton? Is it infinite, as is normally assumed when we envisage a stable proton?

(x) The consistency of cosmological considerations require there to be very much more matter than the luminous matter we can see. What can be said about this 'dark matter' (section 7.4)?

Among the additional problems, the outstanding one is the question of how it all started. One idea is that the vacuum with virtual particle–antiparticle creation is actually unstable in the sense that it can give rise to the real matter and energy in the universe by some kind of fluctuation. These speculations have been further developed by attempting to formulate a wavefunction of the whole universe and then working out from it the probability of material systems appearing from nothing [7.22]. For additional queries, see [7.23]. That science is a search for elusive completeness is really well illustrated by the subject of cosmology. Thus scientific papers and books go rapidly out of date as new phenomena are discovered and new values are found for cosmological parameters.

Of very great interest is also the question of how many advanced civilisations can be expected in a typical galaxy such as our own (the Milky Way). A well-known ('Drake') equation yields a number for this quantity whose unreliability reflects of course the unreliability of the input numbers. It might be five or of the order of millions. If it is a large number you could ask with Enrico Fermi 'Where are they?', meaning the extraterrestrials. If they are more advanced than we are, they should have found a way of making their existence known to us. Or are they so clever that they purposely avoid meeting us? These questions do not belong entirely to science fiction; they are of real concern and of philosophical interest.

There are other, less meaningful, questions which are sometimes raised:

(i) What is the universe expanding into?—The expansion creates the space. It is not pre-existing.

(ii) What happened before the Big Bang?—This is unknown by the meaning of 'Big Bang'. If we can talk about this period at all, we should ask 'What happened 'before the *last* Big Bang' but we are then in an oscillating universe scenario. Some cosmological speculations exist contemplating a universe without initial moment but a finite age and an infinite future.

(iii) Are we at the centre of the universe? Life may possibly exist only on earth! Apart from this important observation, what applies to us should also apply to all other places in the universe, as we might expect from the Copernican principle.

(iv) Why is the universe as big as it is? It is difficult to imagine a very fundamental answer. However, we might observe that life is needed to ask the question, and in the many millions of years required for life to develop the universe has reached a great size.

7.10 Time machines

General relativity (GR) is a successful theory because it is not contradicted by experimental results and gives an attractive account of the effects of gravitation. But it has two severe blemishes.

(i) It allows an infinite density of energy at the Big Bang, since all the matter-cum-energy is then concentrated in a region of zero extension, and something (energy) divided by zero (extension) gives infinity! When a physical quantity can be infinitely large we speak of a *singularity*. Since infinite quantities cannot be measured, or even found, in the real world, the existence of these singularities is a serious problem. Some scientists feel that 'the existence of singularities is an indication that GR is an incomplete theory' and has to be modified [7.24]. I shall comment further on infinities on p 213ff.

While the philosophical aspects of the measurement problem in quantum mechanics have attracted a huge amount of attention (see section 6.6), the same cannot be said of the singularity problem of GR (see, however, [7.25]).

(ii) The GR equations allow for a universe which does not expand or contract, but spins, and in which you can travel backwards in time by the simple device of leaving earth to a great distance, and then returning. It was discovered by Kurt Gödel (see also p 206) and leads to the famous grandmother paradox in which a person travels backwards in time and kills his grandmother so that he himself cannot be born! Note that killing his mother or father early enough

Figure 7.9 A causal anomaly. H G Wells, author of *The Time Machine: An Invention* (London: Heinemann 1895), talking to his younger self [7.26]. Copyright the Estate of Max Beerbohm, reprinted by permission of London Management and Representation Ltd.

in their lifetime will produce the same paradox. This effect plays havoc with causation, for if it really occurs then how can we have a sensible history of anything? GR seems to be too general by allowing singularities and by allowing backward travel in time. Science fiction writers like it of course (figure 7.9). Relativists talk about *closed time-like curves* (CTCs), but the possibility that they might exist is at least controversial.

It is a different matter with forward travel in time, since this cannot affect the historical process by tampering with causality. In fact forward time travel is known from high energy cosmic ray particles which reach sea level by surviving without disintegration far longer than the lifetimes which are observed when they do not travel at great speed. The reason is found in the *time dilatation* effect of special relativity. This tells us that if a person leaves a friend on earth, travels fast for a while, and then returns to the friend, then the traveller will have aged less than the friend. The biological clock is slowed down. This is not at present a practical effect which might enable you to keep young: it requires great speed! It is usually told by reference to twins and is called the *twin paradox*. But let us not call it a paradox here, for it is not a paradox; it is a real phenomenon well confirmed by observations, for example on clocks carried in an aeroplane and on cosmic ray mesons whose lifetime appears lengthened by travelling in space.

Travelling faster than light is another topic which has attracted science fiction writers. It can be contemplated if you use either one of two purely theoretical objects. (i) Particles which always travel faster than light, called tachyons, and which probably do not exist, except as an extrapolation of special relativity theory. (ii) Or you can cut a fast passage by means of an extrapolation of general relativity theory— so-called wormholes. They are not discussed here as being a little too far from standard physics.

There soon appeared papers using ideas of GR to discuss *time machines*, whose main object is to take you back in time, and which are well discussed by one of the authors of these devices [7.27]. This takes one into all sorts of intricate relativistic constructions. I shall not discuss them here, for I believe, along with many other physicists, that history is sacrosanct. This means I believe in a *chronology protection conjecture* [7.28], which suggests that nature is such as to rule out CTCs.

7.11 Summary

I have discussed the ages of various components of the universe. This
has brought in very large numbers which are known only roughly.
The galactic highway code was seen to be governed by Hubble's law
for the expansion of the universe, and I have hinted at the possibility
of a taxonomy of universe models. You can get at it rather simply by
the use of Newtonian cosmology [7.8], which was introduced by W H
McCrea and E A Milne in 1934. I did not do that here, but merely
mentioned the Einstein–de Sitter model, the Milne model, the
steady-state model and the oscillating model. The cosmic background radiation was then seen to limit the acceptable models more
or less to Big Bang types. This was supported further by investi-
gations concerning the abundances of the elements.

The intrinsic incompleteness in cosmology is partly centred on the
beginning, which is beyond normal scientific investigation. The lack
of knowledge type of incompleteness has also been shown in the present exposition: we do not even know Hubble's constant yet with
good precision, and we do not know which is the best model for our
universe, though the choice has been restricted a great deal over the
last ten years or so. The eventual heat death of an ever-expanding
model also requires attention, as entropy can be generated for a very
long time until the model universe remains at roughly constant
entropy and so becomes 'adiabatic and hence dull and lifeless' [7.29].

The problem of how to combine quantum mechanics and the theory
of general relativity into a single convincing theory, called quantum
gravity, is one of the great problems of physics. Why? Because the
replacement of two theories by a single over-arching one has always
had remarkable implications: Maxwell's success in combining theories of electricity and magnetism gave us radio, television and much
else. Newton combined local and astronomical mechanics, and that
yielded the impressive structure of Newtonian cosmology.

We have now covered the very small, the very large and the intermediate size. There are some pretty important matters left: numbers,

God and human aspects of science. They will occupy us in the last three chapters.

(Answer to the problem in section 7.3: yes; for example models with a positive cosmological constant. An accelerating expansion has been mooted recently.)

General relativity interprets gravitational effects as due to curvature of space. An analogy sometimes used is to compare space (at least in two dimensions) with an undulating sheet. On it a ball will tend to roll towards a depression in the sheet. Thus sheet curvature mimics the effect of a gravitational force in a very rough manner. In fact cosmological studies currently suggest, by a rather indirect interpretation of experiments, that 'space is close to flat' [7.30] after all.

As regards time, note that when we see the most distant object visible with the naked eye (the Andromeda galaxy) we see it as it was two million years ago by our reckoning.

Chapter 8

Weirdness or purity
Mathematics: science as numbers

8.1 Introduction

A tool for thought has emerged from our discussions. Look hard for regions of incompleteness in science and, when they have been discovered, look deeply into them since new insights are then likely. This principle is obvious to many scientists.

You may say: this is a principle of science; it does not apply to mathematics, since mathematics has been made by man himself, so that everything in it must be complete and clear. However, in trying to understand nature, man applies to the phenomena he sees around him a kind of grid of knowledge, consisting of man-made concepts and logic, and this grid never fits exactly. It can be made finer, thus reducing the approximations involved. However, there must always remain a mismatch. Completeness of fit eludes. Now things may be expected to be different with man-made ideas: our own inventions, using clear definitions, present us with a different situation. At first sight, therefore, my principle does not seem useful in this context. We shall see in the present chapter that this view is incorrect.

Before we raise more question marks about our knowledge and about science, let us realize that there exists at least one true proposition. This will be established by dry logic! Consider two propositions:

A. There is at least one true proposition; B. A is false.

If B holds, A is false and there are no true propositions. Therefore B, being a proposition, cannot hold either, i.e. it is false. Therefore A is true, for only then are we free of contradictions. It follows that there must be at least one true proposition.

8.2   Gödel's theorem: consistency and incompleteness

A famous result of the 1930s deals with mathematical ideas which depend on *axioms* and deductions from them. They are called 'formalized' mathematical theorems. That there is trouble with such theories was first suggested by some innocuous puzzles. Think for example of Groucho Marx, who said that he would not think of belonging to a club that was willing to have him as a member. That is amusing, but it raises no logical difficulty, since it is equivalent to saying that he will not join *any* club.

More serious is the case of the Cretan philosopher Epimenides. He is reputed to have lived in the 6th century before Christ in the city of Knossos on the island of Crete. He is supposed to have lived for several hundred years and to have been asleep in a cave for 57 of these. It is hardly surprising that after his death the Athenians pronounced him a god. He, a Cretan, said that all Cretans are liars, and if it is applied to all statements and to all Cretans, then this statement itself is a lie. So it is *not* true that all Cretans are liars; it follows that some Cretans speak the truth some of the time. In this argument we have deduced from one proposition its negation. This is called a *paradox*—a statement which is contrary to received opinion, conflicts with preconceived notions, or harbours a contradiction. This type of paradox launched the problems to be discussed in this section.

I will not bore you with related examples like 'Every rule has an exception', 'Never say "never"', etc. If you think these through, you will find paradoxes. They are due to the fact that these statements contain a self-reference for they are statements about statements. The study of these paradoxes found their culmination in results discovered by the Austrian logician Kurt Gödel (1906–1978), the Englishman Alan Turing (1912–1954) and others.

When certain shortcomings in mathematics were found along lines analogous to the liar paradox, we looked for greater rigour in mathematical proofs. Just as in Euclidean geometry, we wanted to be clear about the basic assumptions (axioms) and the results deduced from them by logic. These are the *propositions*. Important propositions were called *theorems*. So now *formalized* mathematical theories were developed, i.e. theories which relied on axioms and the propositions deduced from them, as already familiar from Euclidean geometry. For our purposes the numbers 0, 1, 2, ... will also be considered part of such theories. The distinguished German mathematician David Hilbert (1862–1943) had proposed exactly this programme for mathematics: theorems were to be logical deductions from a set of axioms. It was this plan which was demolished by Gödel's theorem to be discussed below.

While talking about rigour, will what is regarded as rigorous *now* still be regarded as rigorous by mathematicians in 50 years time? It is hoped that this will be so, though nobody can really be sure of it.

Let us talk a little about typical propositions, like the statements I have made above in inverted commas. Let us call them A, B, etc. For some quite specific formula A in such a theory we may perhaps prove either that it is true, or that it is not true. Then all is well, and if we can do this for *all* formulae, then that particular formalized theory is in good shape and is said to be *complete*.

Of course the theory would be in bad shape, and must be discarded, if it contained *even one* formula A such that it is possible to prove both 'A' and 'not-A', i.e. that A is true *and* that A is false. Such theories are called *inconsistent*. An inconsistent theory must be rejected simply because, if it is accepted, then you can deduce *any* proposition from it. Russell is reputed to have made a remark to this effect at a dinner party. When pressed to explain, the story goes, he asked for an inconsistent proposition, and was offered '2 = 1'. 'All right', said Russell, 'what do you want me to prove?' 'Well', came the answer, 'show me that you are the Pope.' 'The answer is simple', remarked Russell. 'The Pope and I are two people, but as 2 = 1, so the Pope and I are one.'

Between these extremes of complete formalized mathematical theories and inconsistent ones there lie *consistent but incomplete* theories.

If one exists, it contains at least one undecidable proposition, A, for which it is not possible *either* to prove 'A' *or* to prove 'not-A' within the rules of the theory. It turns out that this class includes all formalized mathematical theories which contain the normal arithmetic of numbers, and therefore also most formalized scientific theories. It is my job to explain this surprising implication of what is called Gödel's first theorem. It shows that there are limits to what can be achieved by formal reasoning. Many key mathematical problems are now known to be unsolvable. Gödel's second theorem shows that a mathematical theory cannot contain a proof of its own consistency [8.1–8.4]. As these things are complicated, let us leave generalities and look at an example.

Let W be an adjective like 'English' or 'French' and consider for our theory T the statement S:

$$S: ` `W' \text{ is } W$$

For a given adjective W this statement is true, false or meaningless. If W is the word 'pretty', then S can be regarded as meaningless. But for simplicity I shall omit the possibility of it being meaningless, as this does not affect the basic result [8.5]. If the proposition S is true, I shall say that W is 'autological'; if false, 'heterological'. Thus the word 'English' is autological, the word 'French' is heterological, but 'francais' is again autological. A complete list of standard adjectives is assumed given, but we have now extended the list by two new and artificial ones. This raises the question whether the statement S applies to them.

Consider 'heterological'. If S holds for it then 'heterological' is autological. Thus it applies to itself. So it is heterological, which is a contradiction. If S does not hold for it, then S is false, and 'heterological' is by our definition heterological. So it applies to itself! So it is autological, and we again have a contradiction! Neither S nor its negation can hold for the case when W stands for 'heterological'. We have found one statement which cannot be proved or disproved. So the system is indeed incomplete. Although this example is drawn from logic, it is similar in structure to appropriate examples drawn from mathematics.

The word 'autological' does not lead to a paradox; it is clearly autological. But is it? Suppose it does *not* apply to itself. Then 'autological' is heterological and this, again, does not lead to a paradox. Remarkably enough, 'autological' can be both autological *and* heterological without contradiction.

We might be able to get out of the difficulty of not being able to prove a certain proposition which is believed to be true, by going to some larger consistent logical system. But the same sort of problem will arise again: it will be possible to show that the larger system is incomplete with respect to some new proposition. The matter cannot be settled. Thus in every consistent formal mathematical theory containing arithmetic, there will be found propositions which are believed to be true, but which cannot be proved within it, as already noted. We do not need a statement of Gödel's theorem here. But note that one consequence of it is this: in every consistent formalization there will be arithmetic truths which cannot be proved within the mathematical system, i.e. by using only the rules of this system; but they can be established by going outside the system.

This opens up an astonishing gap between what, on the one hand, is provable in the larger system and what, on the other hand, cannot be proved in the smaller system. It reminds us of one of the purposes of this book in drawing attention to our intellectual limitations.

Now to some implications of Gödel's theorem. A hand-held calculator reminds us of the mechanical nature of human computations. We are given numbers, we are given rules for their manipulation, and hence the calculation can be made mechanically. Chess machines as well as giant computers are very complicated. A basic machine, an abstract model of a computer was introduced by Turing. As far as I know a model of it is not even marketed—it would work too slowly.

Let us consider the *Turing machine*. It is started by giving it a paper tape containing the program followed by a blank space and then the data on which the program is supposed to act. Theorems and formal systems generally can be manipulated by a Turing machine: it is a theorem-proving machine. Thus what can be achieved by manipulation of formal systems in a normal computer can also be achieved by Turing machines, which should also be able to prove Gödel's theorem. Of course this would take a long time.

The paper tape is divided into square cells each of which is either blank or has a 'one' printed on it. The machine moves the tape to the left or right, one square at a time, and thereby puts a 'one' in a square, erases a 'one', or leaves the square blank, depending on the program that has been fed in. The totality of these moves is a 'calculation'. If there is an end to it, the machine turns itself off—it 'halts'. So we now have a marked paper tape which is the solution to the problem posed.

A machine can get itself into a loop of instructions so that it will never stop, and it can be hard to know beforehand if that is what will happen. The nature of the calculation is determined by a 'program', which is a list of 'instructions'. Another 'insolubility theorem' is in fact the so-called Turing's theorem. It says: given a Turing machine and a program, there is no standard and general way of predicting if the machine will ever stop running. Of course, you can run the program and wait. But how long should you wait? The *halting problem* presents precisely this rather hard question.

The Gödel and Turing theorems show that in mathematics, just as in science, hard limits are set to what we can know. What is helpful, though, is that at least we can get to know what these limits are.

Whether, or not, the human mind can derive results that go beyond what can be proved by Turing machines is the subject of a long-standing discussion. The common-sense guess would be that certainly our consciousness enables us to go beyond any mere computational activity, in agreement with what has been argued in much detail [8.6], the discussion having been initiated by Turing's own thoughts of the 1930s [8.7].

8.3   Complexity and randomness

Consider the ratio of the circumference to the diameter of a circle. It is the same for *all* circles, and has been denoted by $\pi = 3.141\ 59....$ It is a perfectly well-known number, but, written as a decimal number as above, it is never ending. It is not even known if there is a repetition of a whole string of numbers or if the numbers form a random sequence. Here we have another limit to our knowledge.

Now a typical tape for a Turing machine is just a jumble of zeros and ones, corresponding to unmarked and marked squares respectively. This might result from the use of a simple numerical code, the beginning of which is shown in table 8.1.

Table 8.1 A binary code.

0	1	2	3	4	5	6	7	8	9
0000	0001	0010	0011	0100	0101	0110	0111	1000	1001

One merit of this binary code may be explained by supposing that someone has thought of one of the above ten numbers, and that you want to find it by the least number of questions to which the answers are allowed to be only 'yes' or 'no'. Four such questions turn out to be enough. The simplest procedure is to divide the interval as nearly as possible in two halves, and keep doing it. Thus, ask:' Is the number bigger than four?' If the answer is 'no', you ask about the interval 0 to 2, or 2 to 4. If the answer is 'yes', you ask about the interval 5 to 7 or 7 to 10, etc. This procedure is much simpler, and more obvious, if you imagine that the relevant numbers are written in the binary notation of figure 8.1. All you need to do now is to ask: 'is the first number a 1?' and proceed to 'is the second number a 1?', etc. It is now obvious that a maximum of four questions is sufficient to produce the answer.

Incidentally, the answer to a yes–no question of the above type extracts what is called a *binary digit* or a *bit* (for short) of information. On average, printed English contains 10 bits per word, or 1.9 bits per letter, as may be seen by the following language game. Think of a common non-standard sentence, and challenge someone to guess letter by letter what sentence you have in mind, counting blanks between words as letters. The answers must be 'yes' or 'no'. When the sentence has been determined by the guesser, the number of questions asked, divided by the number of letters, is the information in bits per letter contained in the language. Owing to the existence of spelling rules, standard endings, etc, the result obtained is much less than it would be for a hypothetical language *without* spelling rules and containing 27 equally likely letters. You would then find 4.7 bits per letter by using some simple formulae, rather then by experiment. The theory used here is called *information theory*.

Of course many codes are known, and some encode texts rather than numbers. So if you are presented with a long set of zeros and ones, it may be unduly specific to call it a 'number'. It is better to call it a 'string' for it could represent a number *or* a text. An interesting question to raise about such a string is to ask how *complicated* it is. You could also ask if a given theory, whose results for a certain specific situation are offered in the form of a string, is complicated. This raises what is technically called the question of 'complexity'. It has been suggested, notably by Gregory Chaitin of the IBM Watson Research Centre in New York State, but also by others, that the length of the shortest computer program which could generate the string can serve as a measure of its *complexity*. This length can be measured by the number of operations needed to type the program. Similarly, the complexity of a number can be taken to be the complexity of the shortest program that generates that number [8.8].

This is a philosophically interesting step for two reasons. First, it gives us a criterion for a random number. We can now define a random number as one such that there is no program for calculating it which is shorter in length than the number itself. Thus randomness implies non-compressibility of the appropriate program or number. It has been suggested incidentally that most numbers are random. Secondly, we are reminded by these thoughts that the whole idea of a scientific theory is to encapsulate what is known in the briefest possible way: knowledge represents a reduction in complexity.

Box 8.1 Numbers in words?

I shall end this section with a simple warning that there are logical traps in the description of numbers by words. Should such attempts be forbidden by law? I am thinking of the Pythagoreans, who discovered that if the side of a square is five units long, then the length of the diagonal cannot be given as a simple number of similar units; nor can a ratio of integral numbers of such units give one the length of the diagonal. Further, this holds for *any* square. This seemed most embarrassing to the Pythagoreans who believed in the importance of integers. It was to be kept a secret under threat of death! Leopold Kronecker (1823–1891), many centuries later, voiced a related view when he said that the integers are made by God, the rest is the work of man.

Consider next the description of numbers in words. First example: the number 111 777, i.e.

one hundred and eleven thousand seven hundred and seventy seven

It is described by 19 syllables. You can easily convince yourself that, of the integers that need 19 syllables or more to describe them in English, it is the smallest integer. In fact, it is

X: the least integer not nameable in fewer than nineteen syllables.

But this sentence has 18 syllables. You arrive at the result that the least integer not nameable in fewer than 19 syllables can be expressed in 18 syllables. This is again a simple paradox.

Second example:

Y: let us call N the smallest number which cannot be expressed in words. But, look, we have just expressed this number in words. This is another simple paradox.

The resolution of these two paradoxes is that the numbers specified by X and Y cannot actually exist. In mathematician's language: the sets involved are empty.

8.4 Infinities

In mathematics 'infinity' has the interpretation of something larger than any conceivable number. For example, the numbers 0, 1, 2,... go on and on, and we shall call their number *aleph-zero*. This is also called a 'denumerable infinity'. It expresses the notion that while you can count the numbers, there is an infinity of them. The number of points on a line is also infinite in number, but they cannot be counted; so they are 'non-denumerable'. Their number is said to be 'aleph-one', but we will not require this number here. Just note that this non-denumerable infinity is as large for a short line as it is for a longer line. The Hebrew letter aleph is normally used in this context.

The well-defined, but surprising, properties of aleph-zero were illustrated by Hilbert as follows. A hotel had many (aleph-zero) rooms and many (aleph-zero) guests, each guest occupying one room. The hotel was full. When an additional guest arrived, however, the management was able to accommodate him. The new guest was placed into room 0, while the occupant of room 0 was moved to room 1, the occupant of room 1 was moved to room 2, and so on. So aleph-zero rooms were enough. Indeed Hilbert's hotel can accommodate not only aleph-zero guests, but even two groups of guests, each aleph-zero in number. The first group is simply assigned the even-numbered rooms, of which there are aleph-zero, and the second group is assigned the odd-numbered rooms, of which there are also aleph-zero!

In 1921 Hilbert is also supposed to have said (presumably in German) 'Infinity! No other question has ever moved so profoundly the spirit of man'. In fact, it is simple to handle certain cases of infinity. For example, you can add one half to one quarter, and then add one eighth, and carry on with this addition to infinity. Each time you add half the number you had last time. The sum turns out to be one! In this case infinity need not frighten anybody!

There are other cases when you slip into some recurrent situation which goes on for ever. Here is an example. Take a three-digit number whose digits are not all the same. Arrange the three numbers in order of decreasing size. Subtract the number which results if you arrange the digits in order of increasing size. Repeat. You will always end up with the number 495. After that you are stuck at this number since 954 − 459 is again 495. This is a kind of *arithmetical black hole!* Another example arises if you divide one by three. The decimal number which results is 0.3333... the sequence of threes does not terminate. But this number can be written equivalently simply as 'one third' or 1/3 .

That was 'only' mathematics. Turn next to a famous Laurence Sterne novel in which *Tristram Shandy* writes his autobiography, but covers only one day of his life in each year of writing. This has led to a discussion of infinity and its meaning, and whether or not an infinite past is possible or not (see for example, [8.9]). Both Kant and Bertrand Russell contributed. Here our interpretation is simply that the autobiography will never be completed!

But if infinities arise in science then our suspicions are aroused. Was there an error in the underlying theory—for infinities can certainly not be measured? Perhaps an excessive idealization gave rise to it? Or was it some clumsiness in the choice of variables? In any case, the basis of our understanding must be regarded as somewhat uncertain.

Let us go back to Chapter two and the unattainable absolute zero of temperature. Since we cannot ever measure infinite quantities, how about changing the normal absolute temperature scale, denoted by T, to another one: $T' = 1/T$. The temperatures on the new scale are written below the corresponding ones on the old scale.

T	0	0.1	1	10	100	1000
T'		10	1	0.1	0.01	0.001

We see that on the new scale the lowest absolute temperatures are very large, and absolute zero becomes infinitely large. This is quite sensible, as we would expect an infinite temperature to be immeasurable. This is one way of interpreting infinities in science: they are unattainable [8.10].

Let me mention some other supposed infinities. The equivalent temperature of the simplest type of black hole increases as its mass goes down. These black holes are believed to evaporate. So what do you expect? As the mass goes to zero, its temperature becomes infinitely high. This does not make sense and has led to the suggestion that the object ceases to be a black hole when it is very small, so that in the neighbourhood of these conditions there is a lower limit to the possible black hole mass.

Again, if you really try in your mind to approach the Big Bang, you often suppose that the energy density was then infinite: all the energy was concentrated at a point. But this is not sensible, even though we sometimes talk like this:

> *'According to the standard big-bang theory the universe came into existence in a moment of infinite temperature and density some fifteen billion years ago.'* ([8.11], p 138)

> *'...there was a moment some ten to twenty billion years in the past when the temperature of the universe was infinite.'* ([8.11], p 158)

(The American billion used here represents a thousand million.)

> *'...in my favourite way of looking at this, there is an infinite number of them (universes) and this number is constant.'* ([8.12], p 85)

We have to accept that current models cannot be expected to apply at this early stage, and that these very early times are still poorly understood. In fact, the occurrence of an infinity in science is in my view a warning signal that something has gone wrong or is not fully understood.

8.5 The physical constants

A distillation of the results of physical science is represented by the so-called fundamental constants. We have already met several of these (table 8.2).

There are two more problems with these numbers. (i) How do we account for their values? It might, after all, be possible to proceed in a manner analogous to the calculation of the number π. That number was calculated by mathematicians and the result compared with actual measurements. This is possible since the diameter of a circle and the length of its circumference are related by the number

Table 8.2 The fundamental constants and their effects
(the usual symbols are given, but not the numerical values).

Planck's constant	gives the uncertainty principle (pp 46, 124)
The speed of light in vacuo (c)	maximum speed of energy (p 177)
The electron charge (e)	determines electron properties (p 50)
The electron mass (m_e)	determines electron properties (p 41)
The proton mass (m_p)	determines proton properties (pp 41, 64)
Newton's gravitational constant (G)	determines the gravitational field (p 161)
The current value of the Hubble parameter (H)	determines the expansion of universe (p 175)

Table 8.3 Dimensionless combinations of constants.

a	Fine structure constant $e^2/2\,\pi hc = 1/137$
b	Ratio of proton to electron mass $m_p/m_e = 1836$
f	Ratio of the electric to the gravitational force between a proton and an electron at a given distance apart $e^2/Gm_p\,m_e$ = a 40 digit number.
d	Ratio of the size of the observable universe to atomic size $m_e\,c^3/H\,e^2$ = a 40 digit number

Both f and d are so large as to be hard to visualize. They even exceed the number of cells (ten million, a seven-digit number) the human body is believed to lose every second; or the 14 digit number which would give the loss in a year.

π (=3.1416...). Can we proceed similarly with the numbers of table 8.2? Unfortunately, there is no accepted way of doing so. Indeed some scientists would say that this is not a worthwhile problem at all. (ii) Are these number really constants, i.e. independent of time?

As regards (i), we are on the shore of the unknown. All attempts to explain the values of these constant have been failures. Perhaps the best-known attempt was due to Arthur Eddington (1882–1944) who made determined efforts in this direction, which were summarized in a posthumously published book [8.13]. All constants came out more or less in agreement with experiment as then known. It was so tightly argued that any error in one place would throw things out practically everywhere else. The arguments were hard or impossible to follow.

I was keen to understand something so fundamental when the book came out, and wrote (as a recent PhD graduate!) to the distinguished Professor of Mathematics at the University of Edinburgh, Sir Edmund Whittaker (box 8.2), when I found an error early on in the book. For the reasons I have given this was a rather serious matter, and figure 8.1 shows his reply. I was disappointed by all this, first because I was a great admirer of Eddington, and second, because my request for information as to where I could find the 'alternative derivations' was never answered. I regretfully concluded that this approach did not work, and this has been substantiated by subsequent writers.

Box 8.2 E T Whittaker.

Edmund Taylor Whittaker (1873–1956) was a distinguished mathematician, whose first book, *A Course of Modern Analysis*, was published in 1902 and became a standard work. Many other books followed, notably his *History of the Theories of Aether and Electricity* (1910). Its revised and enlarged edition was completed in the early 1950s, when he was 80 years old, and this two-volume work became a classic. He wrote many other books and papers. The postulates of impotence were emphasized in *From Euclid to Eddington* (1949 Cambridge University Press). He was very active in the Royal Society of Edinburgh, of which he was President 1939–1943, and held the Edinburgh Mathematics Chair for 34 years.

Postulates of impotence are not the direct result of measurements, but express a conviction that all attempts to do a certain thing are bound to fail. He thought that one day a treatise in any branch of physics could be written by deriving everything needed by syllogistic reasoning from postulates of impotence. Though Sir Arthur Eddington held somewhat similar views, they are not now widely supported.

---

from Sir Edmund Whittaker, 48 George Square, Edinburgh, 8.

1950 April 2.

Dear Dr Landsberg

When Professors Temple and Copson and I were reading the proofs of 'Fundamental Theory', we noticed the difficulty you mention, and spent a lot of time in trying to clear it up. But in the end we decided that at this point, Eddington was just wrong.

The error affects the proofs of several theorems after it, but happily the rot doesn't extend very far, as an alternative derivation can be found for some of the later work.

Yours sincerely

E. T. Whittaker

Figure 8.1 A letter from E T Whittaker.

As regards (ii), there have been suggestions that Newton's gravitational constant might decrease with time. This idea was originated by Dirac in 1937 on grounds which I shall now explain.

Units such as metres, grams, seconds, watts and so on, in which the various constants are measured, are of course invented by scientists, so that the numbers that go with the items in table 8.2 contain a strong human element. In order to eliminate this, we must consider ratios such as those of table 8.3. They have the same numerical value whatever units are used, so that they can be regarded as more fundamental. (There are of course other examples in which the units cancel out, again giving values which are independent of units.) Dirac drew attention to the dimensionless ratios f and d of table 8.3 in 1937.

One sees that some ratios, a and b in table 8.3, are 'of order unity'; note that 1000 and 1/1000 still count as 'of order unity' in the context of these very much larger numbers. Other ratios are incredibly large: numbers f and d in table 8.3. Dirac argued that such large numbers cannot be expected, so that if two such numbers do occur, their similar size must be of fundamental significance. If so, they must *always* be of this size. However, number d increases with time since the reciprocal of Hubble's parameter (the *Hubble time*, figure 7.3) represents a rough estimate of the age of the universe. It then follows that, if numbers f and d are always equal, then number f must also be expected to increase with time. The only one of its components which can be expected to change with time is G. From this he inferred that the gravitational constant must slowly decrease as the universe ages. We cannot easily attribute this effect to any specific mechanism, so that we may need a Dirac demon (see p 78 , [8.14]) to bring it about.

Many attempts have been made to try and find this effect. However, G is hard to measure. The accuracy of its measurement is much worse than for the other constants. They are known to one part in a million or better. So all we can say is that the rate of change of G is pretty small: it is less than about one part in a million million of G itself, every year [8.15]. This does not mean that G *is* constant in time, merely that it *might* well be. Thus Dirac's *large number hypothesis* (LNH) remains in the subconsciousness of physicists as probably of marginal interest, but ready to be unearthed should the occasion arise.

I do not want to give the impression that the above are the only fundamental constants of physics. There are others buried deep in current theoretical physics, but not usually discussed in the present context. They are the coupling parameters of the elementary forces but they are not well known and they lie beyond the present scope [8.16]. They are analogues of the fine structure constant (table 8.3), which refers to the electromagnetic coupling of charged particles.

Eddington (box 8.4) introduced an estimate of the number of particles in the observable universe and gave it in full in one of his popular books (figure 8.2). Being an eighty digit number, it turned out to be approximately the number f or d in table 8.3, multiplied by itself. Only some demon could actually count the number of particles in the universe, and I have called it Eddington's demon [8.14] (p 78).

The limits to our current ability to understand the values of the physical constants is a serious item of ignorance exactly because the values of these constants are so crucial. Most physicists take the obvious way out and regard these constants as simply 'given'. This is fine, but for the purist there is here a deep problem.

CHAPTER XI

THE PHYSICAL UNIVERSE

I

I BELIEVE there are 15,747,724,136,275,002,577,605,653,961,
181,555,468,044,717,914,527,116,709,366,231,425,076,185,
631,031,296 protons in the universe, and the same number of
electrons.

Figure 8.2 Eddington's number [8.17].

8.6 Cosmical coincidences

The fact that two almost equal, but extremely large, numbers occur in
rather different contexts was made use of by Dirac, as we saw in section 8.5. This was a kind of *cosmical coincidence* of two numbers.
There are many more coincidences of this type, though they have not
been utilized to deduce physical effects in the way Dirac utilized his
coincidence.

The simplest approach to an understanding of this state of affairs,
which is rather superficial, is to use *dimensional analysis* (see box
8.3). If we start with Dirac's LNH, most combinations of constants
which represent a mass involve the gravitational constant, G. Hence
they depend on time, unless they involve G only in the ratio G/H. The
only such combination which is therefore independent of time has
been denoted by $M(0)$ in table 8.4. From the expression for M(0) in
terms of the basic constants we can infer an approximate value for
this basic mass, which turns out to be about 0.4 times the rest mass of
the charged pion ([8.18]; see also [8.19, 8.20]). We also find an 81 digit
number for the equivalent number of such particles in the observable
universe, by dividing $M(4)$ of table 8.4 by $M(0)$ for reasons to be
explained below. This is Eddington's number. We see that it has been
obtained here by dimensional analysis using the LNH.

Most of the masses obtained are time dependent. They are denoted by $M(4)$, $M(2)$,...,$M(-2)$ and are shown in figure 8.3. Their interpretations are in table 8.4.

The estimate of the mass $M(4)$ of the universe is obtained by regarding it as a uniform sphere of matter, using the critical Einstein–de Sitter matter density (of p 180). Its radius is taken to be the distance covered by light in the Hubble time. These are reasonable assumptions for such an order of magnitude consideration and justify the above division of $M(4)$ by $M(0)$ to obtain an estimate of the number of particles in the universe.

As regards $M(-2)$, you can argue from the uncertainty relation connecting energy and time. We saw on p 152 that for a short time interval we have a large energy uncertainty. Conversely, this energy uncertainty is minimal for a long time interval, and what longer time can we take than the age of the universe? This smallest energy leads us to expect that the smallest non-zero mass which can be measured is $M(-2)$.

Box 8.3 Dimensional analysis [8.21, 8.22].

The 'dimension' of a distance measured in terms of centimetres, inches, kilometres, etc is denoted by L. The dimension of time measured on terms of seconds, hours, years, etc is denoted by T. The dimension of speed, typically measured in centimetres per second, is then denoted by L/T. From Einstein's equation $E = mc^2$, and denoting the dimension of mass by M, you find that the dimension of energy is $E = ML^2/T^2$. In dimensional analysis you look at equations of physics and ensure that the dimensions are the same on both sides of any equation. You do not always have a precise equation. If the left-hand side is known, but there is some doubt about the right-hand side, then dimensional analysis can often be used to obtain information about that part of the equation.

Box 8.4 Sir Arthur Eddington (1882–1944) [8.23].

There was a famous expedition to Principe Island led by Eddington. It studied the bending of light rays predicted by general relativity ('GR') during an eclipse of the sun on 29 May 1919 (section 2.9.3). The results were favourable to GR and the official verdict was given on 6 November 1919 at a meeting in Burlington House in Piccadilly. The occasion was dramatic, making newspaper headlines the following day. The story goes that J J Thomson was in the Chair and remarked afterwards that he did not really understand GR. After the meeting an early author of a book on relativity, Ludwig Silberstein, went up to Eddington, remarking, 'You, Eddington, are surely one of the three people who understand GR'. Eddington demurred. 'Go on Eddington', said Silberstein, 'don't be modest'. 'I was not being modest', replied Eddington, 'I was just trying to think who the third person might be'.

Eddington is the *hero* of this chapter.

To show how coincidences come about, five masses are listed in table 8.4. They can all be expressed in terms of the constants h, c, G and H of table 8.2, and several of them have a simple interpretation. The results of dimensional analysis are such that the ratio of numbers $M(4)/M(3)$, $M(3)/M(2)$....$M(-1)/M(-2)$ (not all given in the table) is always the same 21 digit number (which is of the order of 100 million million million). It follows, for example, that $M(4)/M(2)$ is the same as $M(1)/M(0) \times M(1)/M(0)$, and you can make up all sorts of other interesting relations [8.24]. These then represent theoretically derived *cosmic coincidences*.

Table 8.4 Some basic masses in grams (shown in figure 8.3).

$M(4)$	a 57 digit number	Mass of the observable universe
$M(2)$	a 16 digit number	Schwarzschild black hole mass whose lifetime is the Hubble time
$M(1)$	1 divided by a 5 digit number	The so-called Planck mass
$M(0)$	1 divided by a 25 digit number	Mass of a pion
$M(-2)$	1 divided by a 66 digit number	Smallest mass which can be measured

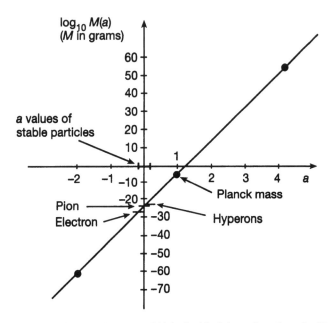

Figure 8.3 Plot of the basic masses $M(a)$ of table 8.4 as a function of a. Note: a 33-digit number = mass of the sun in grams, a 27 digit number = mass of the earth in grams. The vertical axis gives essentially the number of zeros in the mass $M(a)$ when it is expressed in grams. The number in front of the zero is here of little significance. 'a' is a parameter with values from −2 to 4.

8.7 The anthropic principle

The constants of nature must allow for the presence of life on earth. This leads to the so-called

> *weak anthropic principle*: the constants must be such that the universe, when old enough, must have sites where carbon-based life has evolved.

This sensible constraint was proposed by R H Dicke (1916–1997), but it does not represent a fundamental insight like the atomic hypothesis, for example. It is a bit like saying: 'I see a rainbow, it must be raining nearby'.

As an example of its application, note that the weak force (section 3.10.4) determines how much hydrogen is converted into helium from the initial protons in the Big Bang. If the force were weaker, we would have so much helium that the helium content of stars would lead to them burning out rather quickly. There might not then be enough time to form planets and for life to evolve via the complicated nucleosynthesis which yields the carbon which in turn is essential for life. So the incidence of life requires the weak force not to be too weak. There are other analogous constraints on the constants for life to be possible—the abundance of the elements (figure 7.8, p 198), whose explanation is one of the successes of the standard model, would of course be different if the constants were different. For recent surveys, see [8.25–8.27].

A really successful prediction from the weak anthropic principle (below) was made by Fred Hoyle who inferred from the existence of human life and details of nucleosynthesis that there must be a certain energy level in the carbon atom! His argument is too complicated to be given here, but see [8.27].

An implicit use of anthropic ideas is due to Boltzmann. He inferred from the existence of life that our universe may be in a temporal low entropy fluctuation from equilibrium. Life, he argued, can therefore proceed within this lower entropy environment. In it entropy can increase, marking out a direction of time.

A less persuasive variant is the *strong anthropic principle*: the constants must be such as to allow carbon-based life to develop in the universe.

Here you take the physical laws as given, while the constants are regarded as variable from one possible universe to another. The principle itself is a statement of faith which is inspired by the fact that we know of one universe which satisfies this principle. There might be a whole collection of other universes which we do not know about. If they have different sets of constants then life will not be possible in most of them. This is so because, as we have said, the constants have to lie in narrow ranges for life to develop—they have to be 'fine-tuned'. But these speculations violate normal requirements for scientific statements, which stipulate that some experimental check is possible at some time in the future.

So how can we account for the finely tuned universe of ours? (i) Either God created it like that, or (ii) it has to be more or less like this because life is harboured by it, or, (iii) just possibly another intelligence, living in another universe has made ours [8.28]. (iv) Lastly, there is the many-universes theory. This holds that the various possibilities inherent in quantum mechanical experiments are all present in different universes, which, however, find it hard or impossible to communicate with each other. When one of the possibilities occurs here, our universe is supposed to split into several universes in each of which one of the other possibilities is exhibited. On the basis of this model we conclude that our universe is one of many others like it. Life goes on in these as well, but there are differences. A slightly different 'you' may be there for example. Also there may be many more universes in which life cannot exist. This is the fantastic idea of the *multiverse*, but it has support from some respectable physicists (not the present writer!).

Many unresolved issues of this type occur in the scientific enterprise. A decision in favour of one of the four possibilities noted above can be made only by a bold guess, and that is the end of it. You may not even bother to make a private choice. For those that do, it might be the beginning, or the continuation, of their religion: science is here seen as a midwife of guesswork or belief, or, indeed, we may call it faith.

8.8 The Copernican principle

There is a clay model of the world in front of the Hall of Mathematics of the Buddhist temple Yong He Gong in Beijing. It reminds us of the Buddhist view that the centre of the universe is Mount Xumi. But many people follow the Copernican notion (N Copernicus 1473–1543) that not even the earth itself is the centre of the universe; it circles the sun! This idea has been taken further: not only all possible model universes are expected to be realized, leading to the many-universes picture, but also all types of equations with all types of constraints can correspond to universes [8.29].

This *Copernican principle* has been used to estimate the length of life of institutions from the period for which they have already been in existence. If the location in space and time is not in any way special, you can by simple probability considerations infer from the length of its life so far, its likely total life. If you are happy with statements which have 95% probability of being correct, you multiply its length of life so far by 39 to find its maximum total life, and divide it by 39 to find its shortest expected life [8.30]. This has been done with some success in applying the idea to plays and musicals on Broadway in New York and to world leaders' terms of office, including the Conservative government in the United Kingdom [8.31, 8.32].

If you do not want the 95% probability, but a different one, then you have to substitute a different number for 39 above. This can be found by some simple algebra.

8.9 Summary

We have seen in this chapter how the rules of logic, mathematical arguments, the properties of numbers and of physical constants have fascinating implications for our understanding of the universe around us.

Chapter 9

The last question
Does God exist?

9.1 Introduction

This is the age of best sellers. But it would be wrong to suppose that these best sellers are always read. Think of Umberto Eco's *Foucault's Pendulum* or Stephen Hawking's *Brief History of Time*. Ours is actually the age of the unread best seller. In any case, there in Hawking's book we find a remark which can introduce this chapter. He ends his considerations on the origin and fate of the universe by saying that if the universe had a beginning we could suppose that it had a creator. But if it has no beginning or end or boundary then 'it would simply *be*. What place, then, for a creator?' It sounds as if an appropriate and plausible scientific theory can, if we accept it, lead us to dispense with the concept of God. This idea has been much discussed (see, for example, [9.1]). I want to examine it here and, as it turns out, reject it.

In doing so, I shall pretend that you have turned to this chapter first, without having read the rest of the book. The scientific concepts which you will encounter will hopefully encourage you to turn to the earlier chapters for more details.

Here is a brief historical note. The early highly religious period is well characterized by the remark made by Queen Elizabeth I (1533–1603) to a French ambassador: '...there was only one Jesus Christ and one faith, and the rest they disputed about were trifles' [9.2].

A little later there followed the *hero* of this chapter, Blaise Pascal (1623–1662), who helped to found the theory of probability, devisor of an omnibus service in Paris, hailed by the Church, reputed to be connected with a triangle to help find prime numbers, and of course, known for his *Pensées*. He is credited with an argument in favour of ordering your life on the assumption that God exists. For, if you do, and he does exist, then eternal life in heaven will be your reward. If you order your life in that way, and he does not exist, nothing is lost. On the other hand, suppose you order your life on the assumption that he does not exist. Then if he does not exist, again nothing is lost. But beware if he does exist, then eternal life in hell will be your fate, and this is too terrible to contemplate. So what will you choose? Obviously you will order your life on the assumption that he does exist. This is the end of what is essentially *Pascal's wager*. However, I fear that this behaviour will not do you any good, for God will surely see through this manoeuvre, and give you no credit for it!

The above is a classical *game theory* situation, which can be pursued with simple mathematics [9.3]. It then turns out that the more all-forgiving God is, the weaker the selfish rational case for believing in him [9.4]. This is due to the weaker punishment he would then impose on the non-believer. In this generalized context the theory is easily applicable to modern situations whenever punishment and reward are in some competition, i.e. should you, or should you not, attempt to cheat the customs officer?

In the current period of opinion polls, a decline in the belief in a personal god has been noted among the 'greater' (i.e. better known) American scientists, as shown in table 9.1 [9.5]. These numbers, though interesting, do not mean too much as many scientists are religious in a general sense without actually believing in a personal god.

Table 9.1 Belief in a personal god among greater American scientists.

	Percentages of replies		
	1914*	1953	1998
Belief in a personal god	27.7	15.0	7.0
Disbelief in a personal god	52.7	68.0	72.2
Doubt or agnosticism	20.9	17.0	20.8

*Small reductions needed as the figures add up to 101.3%.

9.2 Gödelian statements

The view which I want to defend is actually quite simple: the proposition that God exists can be neither proved nor disproved. The question 'Does God exist?' is therefore regarded by me as an incorrectly posed question. This kind of half-way house is not the opinion of an academic who is unable to make up his mind, or of a philosopher who is forever sitting on the fence. Nor is it in any way a special or unusual situation. On the contrary, it is very frequently true that a proposition 'P' ('God exists', for example) cannot be proved, while the proposition 'not P' ('God does not exist) also cannot be proved. This is so even in the most exacting of all disciplines, namely mathematics (section 8.2).

So we have a whole class of propositions P which can be neither proved nor disproved. In honour of the late Austrian mathematician Kurt Gödel (1906-1978), we can call such propositions *Gödelian*. Following these remarks, I can rephrase my earlier assertion about the existence of God by saying that it is a Gödelian proposition. We should consider here which methods of proof are acceptable— indeed this applies each time a proposition is labelled Gödelian, but we shall not try to do so here.

Some quite distinguished people have classed together the propositions which we may call here non-Gödelian, i.e. those which *can* be either proved or disproved. Schrödinger [9.6] has called them 'trivial'. This is an aggressive word, and I do not like to use it. For example, I do not believe that a beautiful (and easily proved) result such as 'The number of prime numbers is infinite' can be regarded as 'trivial'. But it can be considered as 'beyond discussion'. Indeed we see now that propositions are either Gödelian or (being susceptible of proof or of disproof) beyond discussion. So we come to the conclusion that Gödelian propositions are certainly interesting and worthy of discussion.

We have taken a circuitous route to justify that 'the place of a creator' in discussions of physics is worthy of our attention.

Hints from physics concerning the existence of God have come from thermodynamics, cosmology and quantum mechanics. Please note that I use the word 'hints'. It is a rather cautious term, for you can produce hints in favour of relationships which are actually false. I shall discuss these three cases in turn.

9.3   The evidence of thermodynamics

Thermodynamics is the study of heat and its transformation into work as in internal combustion and other engines. This study must encapsulate the transfer of heat from hot to cold bodies with lapse of time. Put differently, you can measure the lapse of time by looking at how much heat has passed. It has proved possible to sum up this tendency by using the concept of entropy (section 4.1). It is a little abstract because it cannot be measured as directly as can temperature or work. Briefly though, it gives you the tendency of a system to become disordered when left to itself. For example, two bodies at different temperatures, left to themselves and in contact, lose the ability to produce work by means of some kind of heat engine (section 2.6) since they will simply reach equilibrium. This is an example of an increase of entropy and also of disorder. Putting them into contact is rather like *not* utilizing the hydrostatic potential of a waterfall. The employment of a heat engine to produce useful work, on the other hand, corresponds to using the waterfall in a hydroelectric scheme to produce electric power. In the latter case the entropy increase is utilized to produce useful work.

Again, a gas neatly confined to the corner of a box will spread out all over the box if the confining partitions are removed. This is also an example of increase of disorder and also of increase of entropy. Thus it came about that in the mid-19th century Rudolf Clausius was able to announce that the entropy of any isolated system tends to increase. This is one key aspect of the famous second law of thermodynamics (section 2.6).

The second law has led to what we may call an 'entropological proof' for the existence of God [9.7]. This runs roughly as follows. The entropy law ensures that an isolated system reaches internal thermal equilibrium, possibly after a finite time and certainly after a very long time, after which only fluctuations about this equilibrium state can take place. As the universe is far from equilibrium it must have a finite age and hence a beginning. This beginning must be a state of minimum entropy at which the cosmos was born. This was brought about by God, who also created the initial values of parameters such as initial energy, matter, entropy and so on. In this way the universe was wound up, i.e. 'the spring was set'. Gravitation was supposed to keep the process going. Thereafter the universe just runs down to an eventual 'heat death' when everything is at approximately uniform temperature. There are then no stars, as they have burnt out, and there is no life.

This argument assumes that the universe can be treated like an ordinary finite and isolated system so that the second law of thermodynamics can be applied to it. But, granting these assumptions, it is still true that the entropy $S(t)$ at time t could behave in such a manner as to start from zero at the earliest times, and reach some finite positive and constant value at the latest times. The second law is therefore consistent with a universe of finite or infinite age, possibly started off by an act of creation. Thus thermodynamics does not remove the Gödelian aspect of our proposition.

The famous physicist P A M Dirac did not believe in God. So Wolfgang Pauli is supposed to have remarked 'There is no God and Dirac is his prophet'.

The 'proof' of God from entropy is a variant of the third way (of proving the existence of God) given by St Thomas Aquinas. This depends on the fact that everything in the world is *contingent*, i.e. it exists, but it need not exist. Hence there is a reason or cause for its existence. Everything that is contingent has a cause and these causes themselves have a cause. Hence to avoid an infinite regress you may suppose that there is a being who is not contingent but whose existence is necessary. This necessary being is God.

This argument is usually called the *cosmological argument*, or the argument from a contingent being, and has been discussed by many philosophers (Kant, Hume, Russell, etc). As already noted, it does not now rank as a 'proof' [9.8].

In the last century the universe was not usually assumed to have existed for only a finite time (with the resulting problem of a first cause).

It was more often assumed to be static (the matter in the model universe being regarded as smeared out) and to be infinitely old. In that case a thermal equilibrium state would be expected at the present time: no life, no suns, etc. To make an old universe compatible with the facts as we see them (i.e. with life, the sun and stars), one had to assume, with Boltzmann, that it is currently involved in a large fluctuation from equilibrium [9.9]. Again, this does not force us to believe in a creator, but it allows us to do so, if we so wish. This is what emerges in general from our thermodynamic discussion.

9.4   The evidence from cosmology

One hundred years of the second law (1850–1950) and heat death preoccupations was probably enough! With the appearance of more or less popular books on cosmology, for example by Gamow and Bondi in 1952 [9.10], theological interest gradually shifted to the implications of cosmology, to which we must now turn. Cosmology provides many possible model universes.

There now arose a new problem: is our universe ever-expanding, oscillating, in a steady state, or what? The entropy properties of these models became a prime occupation so that even within general relativistic cosmology some excitement about thermodynamics remained unabated.

Although Einstein started modern scientific cosmology in 1917, basing it on the new general theory of relativity, and he had a deep faith in thermodynamics, he could not have known that in the 1970s this elegant theory of general relativity would give birth to the new thermodynamics of black holes (section 7.8), even though no explicitly thermodynamic elements or assumptions had been inserted into the theory of relativity. This is a puzzle which has never been fully cleared up.

The matter distribution in cosmology is often thought of as a 'fluid'. Suppose then that the 'cosmological fluid' in a simple ('Robertson–Walker') model has an an internal energy which depends very simply on the fluid pressure and its volume. Then, remarkably, it can be shown that entropy is conserved in the expansion of the universe. In the case of an oscillation it means that it can proceed in either direction without any dissipation of energy: the model universe could simply retrace its steps while remaining at constant entropy. Indeed, if the universe is modelled by *non-interacting* matter and radiation, its characteristics are symmetrical in time about the state of maximum expansion (see figure 7.4).

Interactions among different particles (waves) are of course crucial (table 3.2) for atomic and subatomic theory. Thus it is clear that we must include interaction between matter and radiation in our model universe. Since the physical conditions are such that the matter temperature is liable to fall more rapidly in expansion than that of radiation, it is easy to see that a thermal interaction between the two must occur: radiation will cool somewhat by giving energy to matter, matter will be heated somewhat by radiation. This will bring about irreversibility. The temperatures are then closer together than they would be without the interaction, and this also affects the pressures of the components. The result is that the model behaves like a thermodynamic system worked irreversibly by external forces. Although there have been doubters, entropy then always increases [7.17, 9.11]. It does so even in the oscillating model.

Relativistic cosmology thus offers a modern interpretation of the heat death should there be an indefinite expansion of the universe. Alternatively, if we live in an oscillating universe, certain aspects of an earlier part of the history of the universe may recur. This is the modern counterpart of Boltzmann's ideas that life is possible because we currently live in a giant fluctuation. Thus there was brought into science the possibility of realizing ancient beliefs in the 'Eternal Return' which the entropy law had previously banished [9.12]. Of course one oscillation is 'easy' to realize as it proceeds from the Big Bang to the Big Crunch when all matter–energy is compressed again. The singularities of relativity theory make it hard, however, to have a series of oscillations. Hard but not impossible, as we can see from various singularity-free models [9.13]. Regarded as unpleasant by Sir Arthur Eddington, and ruled out by the well-known mathematician E T Whittaker, oscillating models are nevertheless again with us:

> *'From a moral standpoint the conception of a cyclic universe, continually running down and continually*

rejuvenating itself, seems to me wholly retrograde. Must Sisyphus for ever roll his stone up the hill only for it to roll down again every time it approaches the top? That was a description of Hell.... It is curious that the doctrine of the running-down of the physical universe is so often looked upon as pessimistic and contrary to the aspirations of religion. Since when has the teaching that 'heaven and earth shall pass away' become ecclesiastically unorthodox?' [9.14]

'The law of increasing entropy definitely excludes the possibility of a cyclic world-process. The universe, then, is running down, and must always run down. Eventually it will attain its state of maximum entropy, when all bodies will be at the same temperature, and all life will have ended. Since entropy is essentially positive, its steady increase must have had a beginning—a creation—when the total entropy of the universe was less than it has ever been subsequently. It was never possible to oppose seriously the dogma of the creation except by maintaining that the world has existed from all eternity... The knowledge that the world has been created in time, and will ultimately die, is of primary importance for metaphysics and theology: for it implies that God is not Nature, and Nature is not God.... For if God were bound up with the world, it would be necessary for God to be born and to perish.' [9.15]

This was written with great conviction by an important authority. But its scientific content is now considered to be flawed. 'Look hard' says our principle of incompleteness (see section 1.4), 'and incompleteness will soon be seen to raise its head.'

There are other models which are not oscillating and also have no beginning or end in time. We note first that the steady-state theory (p 182) flourished from 1948 until the discovery in 1965 of the 2.7 K background radiation (p 184). It requires no beginning and no end to the universe. Then there is a model proposed by Hawking and collaborators:

'So long as the universe had a beginning, we could suppose it had a creator. But if the universe is really completely

self-contained, having no boundary or edge, it would have neither beginning nor end: it would simply be. What place, then, for a creator? .... There would be no singularities at which the laws of science broke down and no edge of space–time at which one would have to appeal to God or some new law to set the boundary conditions for space-time. One could say: 'The boundary condition of the universe is that it has no boundary'. The universe would be completely self-contained and not affected by anything outside itself. It would neither be created nor destroyed.'[9.16]

Do these cosmological models give a pointer to God? The steady-state universe, the Hawking model and the infinitely oscillating model decidedly do not. We might almost regard them as models manufactured for a Society of Atheists. Ever-expanding models, or models with a finite number of oscillations, pose the question of a beginning. If the universe did not develop by itself, then it was created, and God can come in as creator. But this does not clinch the matter since even in the atheists' models believers may well see God's hand in providing the push to move the model along, and in the design of the subtle laws of science. You simply cannot disprove God!

Modern cosmology has given us new concepts and mechanisms for the early universe. The 'age of the universe' at least in the sense of 'the time since the last Big Bang' is *now* a respectable concept (section 7.1), which Chemistry Nobel Laureate Walther Nernst still regarded in 1938 as unscientific, as did Hoyle after him.

Is there such a big difference between believing (a) in an uncaused universe and believing (b) in an uncaused creator? On this question of the existence or otherwise of God, cosmology seems to throw little light. Indeed the attempt in 1951 by Pope Pius XII to look forward to a time when creation would be established by science, was resented by several physicists, notably by George Gamow and even George Lemaitre, a member of the Pontifical Academy. Both were well-known early cosmologists. Some people seem to have felt that there was a struggle between a Christian–Judaic Big Bang and a communist–atheist steady-state model of the universe! Fortunately Pope John Paul II was (in 1988) more cautious:

'... some theologians, at least, should be sufficiently well-versed in the sciences to make authentic and creative use of the resources that the best-established theories may offer them. Since an expertise would prevent them from making uncritical and overhasty use for apologetic purposes of such recent theories as that of the Big Bang in cosmology. Yet it would equally keep them from discounting altogether the potential relevance of such theories to the deepening of understanding in traditional areas of theological enquiry.' [9.17]

The inference 'Big Bang, therefore God' is clearly not acceptable. (On top of this, a future revised cosmology might then bring down God along with the Big Bang with a big bang). The converse does not get us very far, but at least it is more acceptable. It runs as follows: 'if there was an act of creation by God, then the universe could be expected to look rather like the current scientific Big Bang model'. But this is no argument for God's existence.

New thermodynamic problems were raised by cosmology. (i) In the early stages of the Big Bang the universe had very high temperatures (table 2.1), so that its material might be imagined to be in equilibrium. How can the later expansion lead to states of still higher entropy? (ii) How does the entropy of the gravitational field affect expansion? (iii) What is the effect of the large entropies of black holes? These questions, however, belong to cosmology and not to theology.

Modern cosmology does not provide new 'proofs', but new terms and concepts. It tells us about what probably went on near the time of the Big Bang and so has provided a better language for theology. It also provides an estimate for the orderliness of the universe. Black holes play an important part in such considerations because of their very large entropy.

This argument works as follows. For the *disorder* (in a technical sense) of a system let us take its entropy divided by the largest entropy it can have under the conditions one assumes. Then a gas of molecules distributed more or less evenly throughout its container is actually in a state of maximum entropy. Therefore its '*disorder*', as defined above, has the maximum value of unity. But we must bear in mind that disorder and entropy do not always behave similarly (see p 112). At the lowest temperatures, when the gas approaches its most orderly state, the 'disorder' is small. If we imagine all the energy-cum-matter of the model universe to be collected into one single huge black hole, then its entropy would be enormous. We can make a very rough estimate of the current entropy of the observable universe, for example by taking into account the background radiation which survives from the Big Bang (see p 184) and pervades all of it. Even allowing for further additions to the present entropy of the universe, which are harder to estimate, we find only a very tiny 'disorder' as defined above [8.14, 9.18]. Its magnitude might be as small as one divided by a million million! Our model universe seems to be very orderly indeed! Thus black holes come in here as a standard of disorder.

Cosmology enriches the background of theological discourse. Conversely, theological considerations can remind forgetful scientists that the very nature of science forces limitations upon it. For example, science can proceed only from effects to causes, or *vice versa*. First causes, however, have a place in theology, but not in science, leaving another big gap.

9.5   The evidence from quantum mechanics

Turning to quantum mechanics, we note that it is a probabilistic theory. It tells us for example the number of (identical) atoms in certain (identical) well-defined initial states that will decay in a given period with, for example, the emission of radiation. It does not tell us when a *given* atom will decay (see p 137). If quantum mechanics were never to be displaced as a basic theory, we would here have an unbridgeable gap, rather like the gap in our knowledge concerning a first cause (if there is one). Hence a God would be a candidate to make the decision concerning the atomic decay (or concerning any equivalent event). However, this idea does not feel right: first this God could be swept away with a possible new theory, more precise than quantum mechanics; second, God's decisions would be subject to the statistical constraints derivable from standard quantum mechanics. Third, the hypothesis that a theory will never be replaced cannot realistically be made.

The reduction of the wave function (section 6.6.2) is another puzzling aspect where God might 'help', but this help would be subject to the same criticisms.

Lastly we come to the proposed, and not yet properly developed, combination of quantum theory and relativity theory, known as quantum gravity. Among the suggestions that have come out of it is the following idea. On estimating the energy in the model universe, including the energy of motion of the swirling clouds of gas and the electrical and magnetic contributions, we find some positive number. But the total energy should also include the gravitational interactions among all the bodies and this is a negative quantity. It can therefore be the case that the total energy of a system with internal motions is actually zero. A good example of this occurs if a body is thrown upwards from the surface of the earth with just sufficient energy to enable it to escape, i.e. with escape velocity (see p 194). This situation is just on the knife-edge between escaping from the earth with speed to spare, and not escaping at all, but falling back to earth. When the system is on this knife-edge, its energy is zero.

Why is the gravitational potential energy negative? This can be thought of as resulting from taking a state with a large separation of all the particles as the standard state of *zero* potential energy. As they come closer together under the action of an attractive force (gravity) the system loses energy and so its energy becomes negative.

Thus we see that the (attractive) gravitational interaction energy among the particles can reduce the total energy of the universe to zero (as in the above case of just reaching escape velocity). Such a zero-energy model universe can emerge from a vacuum state, which is also of approximately zero energy, by a spontaneous fluctuation. This almost fantastic idea does not violate energy conservation and would lend credence to creation *ex nihilo*. It has been discussed quantitatively as a piece of physics within the emerging theory of quantum gravity since 1973 [9.19]. This type of transition at constant energy has been considered in other contexts within quantum mechanics and is called the tunnel effect (see p 138).

The tunnel effect enables a body to pass through a barrier which would be classically impervious to it, simply because the body has classically not enough energy. In quantum theory, however, the uncertainty principle enables the body to 'borrow' the required energy (see p 152).

In order to avoid the problem with the beginning of time physicists have carried out some surprising manoeuvres in their theories, which include the introduction of model universes without any boundary in space or time, and also the use of an imaginary time coordinate. The transition from an imaginary time to normal or real time can be expected to cause serious problems. The position remains, however, that physics is not adapted to handle the problems of the actual beginning. For a detailed discussion, see [9.20].

Some residual questions remain.

The new understanding of cosmology leads to some serious questions for dogmatic religions. It now seems likely that intelligent life exists in many places elsewhere in the cosmos, even though we may not be able to establish contact for the time being. Thus only a believer in a rather abstract God can deal successfully with enquiries which are tied to specific terrestrial occurrences. For how would you answer questions like: did Christ also appear on planet X in the Andromeda nebula? Thus science is here calling for serious refinements in theological thought. As it has done before, it bids serious persons to leave excessively concrete models, pictures and stories of dogmatic theology in favour of more abstract conceptions.

There is of course Fermi's question, noted in section 7.9, that if *extraterrestrial intelligences* (ETI) exist, then why have we not heard from them or even seen them? We feel lonely and neglected! Again much serious material has been produced on this topic [9.21]. Has our galaxy been manufactured by superior intelligences, thus explaining why the conditions on earth are relatively kind to the existence of life—if one compares them with the conditions on Jupiter (say). To find ETI we would have to choose one of the one thousand stars within a distance of one hundred light years from us, which may be the parent of a planetary system. Then you would have to point your radio telescope in the direction of that star and search the appropriate window of radio frequencies. The task is enormous!

Studies have shown that many constants of nature have to lie within very fine limits in order to allow life to develop on earth. This is part of the so-called anthropic principle and it is a technical subject which makes it appear as if the universe is designed most carefully in order to give life a chance to develop in the solar system (section 8.6). That clearly leads to the suggestion that a God designed it all. This case has been argued by H Montefiore, Bishop of Birmingham [9.22]. Trouble resides, however, in a reliance on current science in the sense that the argument will fall to the ground if the relevant science is changed; for example, suppose that the favourable narrow numerical ranges are deduced in the future from a *single* more fundamental physical theory. If that happens, then God has perhaps fashioned the universe so that it can be described by this new theory; but, being just *one* thing, this is less impressive than the thought that he designed the universe by fine-tuning its *many* constants separately in order that life might emerge.

9.6 Conclusion

A very wide spectrum of views has been propagated about our subject. Does physics have room for a creator of the world or not? Let us test what we have learnt by looking at two fairly extreme views due to very learned and respected people. The first, due to E T Whittaker in his Riddell Memorial Lectures, was expressed by him as follows:

> *'When the development of the system of the world is traced backwards by the light of laws of nature, we arrive finally at a moment when that development begins. This is the ultimate point of physical science, the farthest glimpse that we can obtain of the material universe by our natural faculties. There is no ground for supposing that matter... existed before this in an inert condition, and was in some way galvanized into activity at a certain instant: for what could have determined this instant rather than all the other instants of past eternity? It is simpler to postulate a creation ex nihilo, an operation of Divine Will to constitute Nature from nothingness.'* [9.15]

We have seen that the creation ex nihilo can be regarded as a quantum mechanical tunnel effect, so that this is not a clinching argument.

Again the well-known astrophysicist Paul Davies says:

> '*Our conclusion must be that there is no positive scientific evidence for a designer and creator of cosmic order (in the negative entropy sense). Indeed, there is strong expectation that current physical theories will provide a perfectly satisfactory explanation of these features.*' *[9.23]*

I do not believe this full explanation by physical theory is possible. These theories develop, they change, yesterday's orthodoxies become tomorrow's heresies. A detailed contemporary discussion between an atheist (Quentin Smith) and a theist (William Lane Craig) is available for those who wish to follow the intricate details of these philosophical arguments taking account of recent science [9.20].

Questions of how the currently accepted laws of physics got here in the first place, or how it is that there is a universe at all, are not in the province of science. A disproof of God's existence is just not possible. Nor is its proof. This is in agreement with, for example, the views of the late Professor Ayer:

> '*Thus we offer the theist the same comfort as we gave the moralist. His assertions cannot possibly be valid, but they cannot be invalid either.*' *[9.24]*

Thus the proposition concerning God's existence is Gödelian. Using the word 'faith' in a broad sense to stand for prejudice, belief, or opinion, we can say that the question of the existence of God is a matter of faith whose opposite is certainty:

> '*But what about the ordinary man, imbued with insatiable curiosity? He may raise his hand of course and say gravely: 'I want science. Until science speaks I must have no views on controversial topics.' He may well die before the decisive step forward occurs, and some people will want to complete their world picture before they die. The scientific search being incomplete, science cannot help and they are then free to employ faith. Now science will be in some way incomplete at all times in the future, and in the twilight of the scientific Wild West, in the frontier territories of physics, chemistry,*

biology, etc, science must yield to faith. The drive towards completeness is seen most clearly here, where the very elusiveness of the completion makes science give birth to faith. I see science as the midwife of faith. Far from being competitors for the mind of man, science and faith must both take their rightful place if one wants to attempt a picture of the world.'
[9.25]

Chapter 10

Love of my life
Science as human activity

10.1 Happiness

Taking stock, a reader may say: 'I have been introduced to many ideas of physics on which much of science depends, ascended many hills, and reached a kind of mountain peak with quantum mechanics. Here many of the views were startling and fascinating. But the road was hard. Then, via cosmology, physical constants and religion, we reached a terrain, which, while not simple, was at least used by many more walkers. There was still splendour, but things were easier. The tour is now well-nigh completed. Could I extract perhaps some personal benefit, *in addition* to what I have actually learnt?'

This chapter suggests that you can. We have seen from many details what some people have known or suspected for a long time, namely that nature is like a set of Chinese boxes or of Russian dolls, and I have characterized this by talking of the search for elusive completeness. Thus we read in '*Eine Duplik*' by Lessing: 'If the creator offers us a choice between truth and the search for truth, we must reply "Creator, give me the search for truth, for truth itself is for you alone". We continue to search, be it for beauty, understanding, love, ... and some degree of satisfaction is induced by finding at least some crumbs of these on the way to a remoter, but inaccessible, goal.

Of course such a well-known thought cannot be the point of this book. Here the object is to convey the flavour of ideas in *physics*. But (we note again) this book can be read at a second, more philosophical, level, which reminds us that the ends we aim for are often unattainable. Actually, there is a paradox concerning the achievements of ends. We should try *not* to achieve them, for, if we did, the meaning and direction of that part of our life would be lost. But we should also strive to achieve them because of the happiness derived from the striving. This, the 'paradox of the end', can be avoided, for example, by regarding the end as unattainable in the first place [10.1].

The longing for the impossible already occurs in Byron's *Childe Harold's Pilgrimage*:

> *Who with the weight of years would wish to bend,*
> *When Youth itself survives young Love and Joy?*
> *Alas! when mingling souls forget to blend,*
> *Death has but little left him to destroy!*
> *Ah! happy years! once more who would not be a boy?*

Or think of Elizabeth Akers Allen's *Rock me to sleep*:

> *Backwards, turn backwards, O Time in your flight,*
> *Make me a child again just for tonight!*

Imperfection and incompleteness are not only part of our science. The rot extends much further: we are ourselves imperfect by our own standards! We may acquire more skills, more learning, more money, but we must try to live with ourselves—we do not have to be better than others. We must function as we are, and so must have the 'courage to be imperfect' [10.2]. Further, instead of being 'dried up old sticks', it is good to bring passion to bear on our work.

We must not neglect the psychologists and social scientists if we wish to discuss the question of happiness. They have asked people around the industrialized world to reflect upon their happiness leading to actual measures of 'subjective well-being' [10.3]. Among its three components, the one of importance to us here is 'having goals, making progress toward goals and freedom from conflict among one's goals'. The other two components include the ability to adapt to life's changing fortunes and, of course, a positive attitude to life. Omitted, surprisingly, from this list is the satisfaction derived from the exercise of generosity and service to the community. This should also be a subject of study.

Actually, the research into psychological well-being is relatively new, as contrasted with research into physical and material well-being. It is clear from [10.3] that no reasonably reliable clue to a person's happiness is obtained merely from their age, race, sex or income (provided the latter is above a minimum standard).

10.2   Limits of science

Ever since Emil du Bois-Reymond's famous lectures on the limits of our knowledge of nature, this subject has been discussed extensively by philosophers and others. Not in the way done in the preceding pages, as a part of what may be called a science education package, where a conscientious expositor should come across these limits *automatically*. The appreciation of such limits has often been the result of a more direct philosophical approach, leaving full scientific details on one side. These are of course also useful efforts, and Nicholas Rescher in one of his books comes close to the views developed here [10.4]. I would not go with him all the way, however, for example when he says (p 84):

> *'There is no realistic alternative to the supposition that science is wrong—in various ways—and that much of our supposed 'knowledge' of the world is no more than a tissue of plausible error.'*

We are, after all, dealing with (more or less!) steadily improving approximations.

Rescher's book is only one of many by philosophers and scientists, writing popularly, who have discussed this topic, e.g. [10.5–10.9].

Turning to a heavier gun, Einstein reminds us that 'the fact that in science we have to be content with an incomplete picture of the physical universe is not due to the nature of the universe itself, but rather to us' [10.10].

At the opposite pole is what has been called scientism, the idea that science can in principle explain 'everything'. Its advocates, and I am not one of them, suggest that the brain and consciousness will in due course also yield to scientific treatment, as might spiritual and moral questions. These are matters of opinion, representing guesswork about the future, and have been discussed widely. Their mere mention must suffice here to remind us that precisely because of incompleteness and uncertainties in our understanding, we need acts of commitment, i.e. of belief, if we wish to arrive at a full picture of the universe, even if it is only a private picture created for ourselves.

10.3   Distortions: science and the public

The scientific way of comprehending the world is not the only one. The success of science depends to no small extent on the fact that it works within set limits. Thus it does not, *qua* science, deal with the purpose of life, or the purpose of anything else, or with political suppression, or mental disturbance, or starvation, or ruthless exploitation—the list can be extended indefinitely. See a recent strident discussion in which the strange remark, if it is intended to be serious, 'that philosophy is to science as pornography is to sex' is attributed to a former Reith lecturer [10.11]. It is true that science handles cause–effect relations, and does so extremely efficiently. But there is a drawback: first, causes cannot by their nature be within the compass of science. The origin of the universe is not a scientific problem, though those of us who dabble in the theory of model universes may well ask how it—the model—can reasonably be expected to start, or to have started. But it remains only a model. As the Best Man might say of the bridegroom at a wedding: 'He will be a model husband, and, you know, a model is not a real thing at all!'

Recall as an example the remarks of Atkins [10.12] that general relativity 'lets us trace the trajectory of the universe from its birth to its death'. He means of course the model universe, which makes the statement rather less impressive.

Omission of some of these caveats, probably often unintended, leads to distortions of scientific findings. More serious, and I should like to think much rarer, are distortions which verge on the intentional. Research results are more impressive by omitting a cautionary phrase or adding a misleading but publicity-grabbing headline. This may sometimes improve the author's status, his chance to be awarded honours, to be invited to address prestigious meetings, to be granted new research contracts. Such distortions are sometimes possibly in a good cause, but they are distortions nonetheless. A short rule could be: 'inform people about science, do not try and dazzle them with it'.

In order to reduce extremes of unsatisfactory behaviour in public life, a measure of self-regulation has been accepted, certainly in connection with the media; it is also exercised by the scientific community. Papers submitted for publication are reviewed by colleagues and accepted, amended or rejected in response to the comments received. Things slip through the net occasionally—the well-known cases of Sir Cyril Burt's work on intelligence quotients and the Piltdown Man hoax are examples. Straight copying is rare but not unknown, even in theoretical physics (figure 10.1).

Self-regulation is also helped by bibliographical methods. These have led to great volumes of the very useful Science Citation Index (SCI) which tells us annually what papers have been cited, in which publication and by whom. This is helpful to avoid unnecessary duplication of effort. Also it is desirable for authors to find that they are cited by other authors, since it shows that their paper did not sink like a stone in the mass of scientific works. We must remember though that, roughly speaking, half the published papers never get cited at all (though they may of course be read!). This has spawned all sorts of unreliable numerical rules, since it is is easier to count than it is to think! For example in some places scientists are expected to accumulate a certain number of citations, or to publish at least, say, three papers a year. The 'citation record' becomes important: the more you

Euclidean supersymmetry and relativistic two-body systems

1. – Introduction.

Supersymmetry has been intensively used in past years as a theoretical tool in connection with various models of elementary-particle physics. Supersymmetry ideas have also been widely applied in quantum-mechanics starting with the early work of Witten(') as an example of the simplest supersymmetric field theory. At the beginning the motivation for studying supersymmetric quantum mechanics was the discussion of the underyling mechanism responsible for spontaneous SUSY breaking in arbitrary models. Since then the subject has

A non relative supersymmetric two-body equation for scalar and spinor particles

1. Introduction

In recent years, supersymmetry has been intensively used as a theoretical tool in connection with various models of elementary-particle physics. Supersymmetry ideas have also been widely applied in quantum-mechanics starting with the early work of Witten as an example of the simplest supersymmetric field theory [1]. At the beginning the motivation for studying supersymmetric quantum mechanics was the discussion of the underlying mechanism responsible for spontaneous SUSY (break-up) in arbitrary models. Since then the subject has seen continued interest and many articles constructing new and more realistic models have appeared, so that it is well developed by now [2-5]

Figure 10.1 A rare example of straight copying of a scientific paper. The top version was received on 11 August 1987 [10.13], and the bottom version on 28 March 1990 [10.14].

are cited, the better you are supposed to be. But if you make a mistake, lots of people may cite you to point out your error! In this case the citations do not guarantee high quality! Friends may cite each other as a token of friendship, etc. There are many interesting aspects of these problems which cannot, however, be pursued here.

I would like instead to amuse you with a story, presumably apocryphal, about the great American physicist Robert W Wood, who was called in to give some expert testimony in a legal case. To establish his credentials, the judge asked him who was the best American physicist. 'I am' replied Woods without hesitation. Taken aback, the judge probed further: 'Who is the best physicist in the world, then?' 'I am' came the immediate answer. When the investigation was concluded one of Wood's friends took him aside. 'Hi, Robert, you seemed a little conceited in reply to those early questions, did you not?' 'I know' answered Wood 'I felt rather embarrassed about it. But, you know, I *was* on oath!'

Turning to more serious matters, the former Chief Medical Officer in Britain, Sir Kenneth Calman, explained in connection with the BSE† enquiry that British beef was safe, i.e. it was unlikely to lead to a new form of Creutzfeld–Jacob disease. The mother of a son who died from this disease later, was amazed by these remarks, and is reported to have said 'I think most British people thought that "safe" meant "safe to eat". Now Sir Kenneth is saying that beef might have been safe, was fairly safe or possibly safe. He is rewriting the English language' [10.15]. Sir Kenneth explained that 'safe' must mean 'free from unacceptable risk or harm'. Thus we do not mean that a driver we describe as safe will never have an accident. The reader need not be surprised, since he is prepared from what has been said in earlier chapters for the idea that certainty, which escapes us so often even in physics, is bound to be so much more elusive in human affairs.

A natural extension of self-promotion makes some scientists nowadays go to the press or television with their new results, even before they are published. This is regarded as bad practice, since publication is needed as an indication that the work is approved by scientific peers. In the medical sciences there exists the 'Ingelfinger Rule' which some editors adopt. It proscribes the publication of articles whose substance has been reported in the media. (Franz Ingelfinger was the editor of an American medical journal.) There are other stories about physicists being shunned when they were thought to have told the press about their results before going through 'the usual channels' [10.16]. Of course preprint services exist and are useful in making pre-publication results known.

10.4 Science wars?

Some papers in elementary particle physics now have one hundred persons and more as authors, following decisions by a committee. Most of these authors presumably grasp the nature of the experiment described. The list of authors is so long, it cannot be printed in the

† Bovine spongiform encephalopathy.

papers and so it is codified as this or that 'collaboration'. This is a far cry from the great Victorians like Rutherford and Kelvin. Yet it works! Of course sociologists are interested, not perhaps in pursuing the science itself, but in the mechanism of its pursuit, and such collaborations are of interest to them. One of their fathers, Jean-Francois Lyotard, author of *The Postmodern Condition*, died in April 1998, but his ideas have taken some root. They let in the notion that science is a 'language game', dominated for example by information technology and so takes its place among the language games and 'narratives' of politicians and economists. This apparent demotion of science, and of the 'search for truth', opens science up to criticism and comment by sociologists, who judge it of course as just another human activity. This can hurt the pride of scientists, but without having so far any real effect on their work. One day it may even have beneficial effects; perhaps the sociologists can be thought of as analogous to business consultants, who come in to help reshape an organization which, surprisingly enough, was previously unknown to the consultants.

More recently Alan Sokal managed to show up the journal *Social Text* by publishing a spoof physics-type article in it. It had an impressive sounding, but actually nonsensical, title. This caused both amusement and annoyance and led to the suggestion that certain so far respected thinkers could be (unknown to themselves) verging on being impostors. Sokal had in mind French postmodernists like Derrida, Lacan and others and had made fun of them. Both sides of this discussion appear to have now squeezed the last ounce of publicity from it [10.17].

We must admit that there exists in some quarters a human craving for some final authority leading to some final truth and certainty. Since science cannot provide this in any real sense, this has encouraged the development of non-scientific, or even anti-scientific, ideas represented by religious fundamentalism, astrology and the desire to communicate with the dead. This should be met with sensitive understanding rather than by crass rejection on the part of those who do not believe in these ideas. It will also be helped by the effort made in this country and abroad by the movement to promote a better public understanding of science. In addition, a sensible sociology of science [10.18, 10.19] can be stimulating in telling us, hopefully in a novel way, what we scientists are doing!

10.5. Concluding remarks

Typical scientists enjoy their work, and for some it is the love of their life. I cannot promise that they can reach in science an analogue of the love life of, say, Alma Mahler–Gropius–Werfel, but, more modestly, they welcome the complex network of relationships with their colleagues. They create, they respond to the creations of others, they collaborate, and at times they quarrel and often they compete. They know that scientific activity has no end and that the best they can hope for is partial knowledge and normally only minor contributions. This agrees with Newton's remark that he stands on the shoulders of giants. They have consolation in the notion that while only a small number of scientists contribute to real scientific progress, their work cannot be done without the many small contributions of the less visible scientists. This is the Ortega hypothesis (after Ortega y Gasset, the well-known Spanish philosopher [10.20]).

When it comes to old scientific ideas being displaced by new ones, people sometimes refer to Planck's remarks in his autobiography to the effect that old ideas are replaced only when their advocates die, thus making room for new ideas and a new generation of scientists (Planck's principle [10.21]).

Following Thomas Hobbes' *Leviathan*, note that 'Life itself is but Motion, and can never be without Desire' and the Buddha's remark that desire is the cause of all suffering. Desire here hints at presumably so far unfulfilled wishes. Scientists try to fulfil their professional wishes by swimming with the stream of science: sometimes they swim a little faster than the stream, and then have the satisfaction of making major contributions; at other times they swim with it; but quite often they fall behind in the stream. They may then enjoy the sensation of swimming, but what is in the distance they can now not perceive clearly; it appears as a kind of mirage influenced by their own knowledge and past experiences—the mirage seems to be an elusive but beautiful goddess, disappearing in the distant mist of the unknown.

Glossary

Non-zero masses always refer to particles at rest. In many cases more general definitions are available but are not so suitable for our purpose.

absolute zero of temperature (p 11)
The limit to the lowest attainable temperatures. It is denoted by 0 K on the absolute (or Kelvin) scale of temperature.

abundance (of a chemical element) (p 193)
The mass of an element present in a system as a fraction of the total mass present.

æther (p 129)
A medium, whose existence is controversial, and which is imagined to fill all space, allowing light to propagate in it. It corresponds to water for water waves or to air for sound waves.

aleph (p 213)
First letter of the Hebrew alphabet. It furnishes labels to distinguish different orders of infinity in mathematics (by taking the relevant suffix).

angular momentum (p 46)
A quantity related to the work required to set a system into rotation.

anthropic principle (p 224)
It states that the physical constants of nature must be such as to support life. Only a very narrow range of these values is suitable.

anti-diffusion (p 77)
This occurs if, in the absence of external forces, particles (or atoms or molecules) tend to accumulate in a certain volume.

anti-heat conduction (p 77)
This occurs if, in the absence of external forces, heat flows from a cold to a hot region.

antimatter (p 51)
Consists of antiparticles of naturally occurring particles. These have the same mass and spin but opposite electric charge. Uncharged particles can be their own antiparticle (e.g. photons).

anti-thermodynamic (p 77)
Defying the second law of thermodynamics.

arrow of time (p 87)
See Time's arrow.

atomic number (p 32)
Of an element is the number of protons in the nucleus of its atom.

atomic weight (p 31)
The number of times the atom of an isotope of an element is heavier than the atom of hydrogen. For a mixture of isotopes such as occur in normal elements, the atomic weight of the element is obtained by an appropriate average over the isotopes. To achieve accuracy, the basis for the comparison has been shifted to carbon, taken to have atomic weight 12.00.

autocatalysis (p 108)
The phenomenon of a reaction product assisting the reaction to go faster. Thus one of the products is itself a catalyst (q.v.) of the reaction.

axiom (p 207)
A principle adopted for argument's sake.

baryon (p 52, see table 3.2)
A composite particle made of three quarks, gluons and possibly quark–antiquark pairs.

Bell's theorem (p 144)
A theorem used to show that quantum mechanics cannot be interpreted in terms of local hidden variables. In other words, quantum mechanics violates the theorem. Named after J S Bell, who also has other theorems to his name.

beta particles (p 59)
An early name for the electron.

Big Bang (p 180)
A hot Big Bang is normally meant, according to which the universe expanded from a small and intensely dense and hot system.

binary digit or bit (p 211)
Knowledge gained from the answer to a question which tolerates only 'yes' or 'no' as an answer.

black-body radiation (p 149)
Radiation in equilibrium. Its photons have a characteristic energy distribution which can be associated with an absolute temperature.

black hole (p 74, 174)
Matter so dense as to trap even light gravitationally. Can be specified by its mass, charge and angular momentum.

Bohr orbits (p 45)
A classical analogue picture of the electronic motions in atoms. Each 'orbit' is specified by a set of quantum numbers. Since classical analogues were used, this is referred to as the 'old' quantum theory. It was later developed into a replacement of Newtonian mechanics, called 'new' quantum mechanics.

Bose–Einstein condensation (BEC) (p 119)
The accumulation of bosons in the lowest available energy level of a system. Important at low temperatures and connected with superfluidity.

boson (p 47)
A particle (not necessarily an elementary one) with an integer unit of spin. They transmit forces between certain elementary particles. Any number of bosons can occupy the same quantum state (see table 3.2). Named after S N Bose.

bounce (p 191)
The turn-around from contraction to expansion in cosmological models.

boundary conditions (p 97)
The solution of a mathematical problem may depend on the value of the sought-for function either at the initial state, or at the final state or at certain points in space. Such side conditions are called boundary conditions.

brachistrochrone (figure 6.1, p 128)
The path of quickest descent under gravity which joins two given points.

British Association for the Advancement of Science (p 29)
An association of scientists which covers all the sciences. It was used to announce new discoveries, and is now a platform to promote the wider understanding of science. The first meeting took place in York in 1831.

Brownian motion (p 39)
It referred to the continuous agitation of pollen particles in water. Now applied to the permanent agitation of small suspended particles in a fluid. It is attributed to the buffeting of them by larger atoms or molecules and has led to general studies of fluctuations. Named after the Scottish botanist Robert Brown (1773–1858).

Casimir effect (p 152)
Two parallel plates in vacuum tend to be pulled closer together, due to quantum field effects. Named after H B G Casimir (b 1909).

catalyst (this glossary; see enzyme)
A substance which aids a chemical reaction without being changed by it.

chaos (p 94)
Apparently random, but certainly unpredictable, behaviour. It is sometimes seen as a limit of normal predictable physical situations.

charm (p 63)
A kind of quantum number for quarks. See table 3.2.

closed system (p 22)
A system which can exchange energy, but no matter, with its surroundings.

closed time-like curve (CTC) (p 202)
Curve in space–time which would enable an observer following it to move into the past.

coarse-graining (p 83)
This occurs if one regards a whole group of quantum state as one (coarse) quantum state.

coherence (p 131)
An attribute of waves, or parts of waves, whose phase difference (denoted by p in figure G.1) is nearly constant in time. Used in studies of all wave phenomena, e.g. optics.

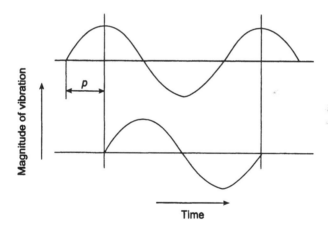

Phase difference p

Figure G.1

collapse of wavefunction (p 125)
See wavefunction.

complementarity, principle of (p 125)
The fact that two modes of description, which are incompatible in the sense of the uncertainty principle, are often needed to fully appreciate a certain situation.

complete mathematical theories (p 207)
Theories in which all propositions can be shown to be either true or false.

complexity (p 212)
There is no single accepted definition as yet. In a physical system it is associated with the ability to participate in a variety of transformations. In mathematics it may be defined by the length of the shortest computer program which can generate the sequence of symbols under consideration.

component (p 157)
See space component.

compound (p 30)
A product of chemical interaction of several elements in definite proportion with resulting properties which differ from those of its components.

conservation of energy (p 17)
Sometimes also called the first law of thermodynamics. Energy transformations in an isolated system leave the total energy unaltered.

conservation laws, e.g. of momentum (p 156)
They refer to quantities which remain unchanged during certain transformations.

consistent mathematical theories (p 207)
Theories devoid of propositions which can be shown to be both true and false.

Cooper pairs (p 163)
Two fermions linked by interactions and thus acting as a boson. Named after L N Cooper.

Copenhagen interpretation of quantum theory (p 125)
There is no accepted definition of this concept. It is understood to tell us that quantum theory speaks of definite states of systems only after a measurement has been made. Before the measurement the state is normally indeterminate.

Copernican principle (p 182)
It states that we occupy no special place in the universe. It has been extended to suggest that we occupy no special position in either space or time. Named after Nicholas Copernicus of Poland (1473–1543).

correspondence principle (p 46)
Mainly due to Bohr, it states that classical concepts of physics have quantum mechanical counterparts. It also suggests how we can pass from quantum to classical results.

cosmic background radiation (CMB) (p 184)
A uniform flux of microwave radiation believed to permeate all space and to be a remnant of the radiation emitted during the Big Bang.

cosmic rays (p 57)
Particles entering the earth's atmosphere from all directions with relativistic speeds (q.v.). They may be emitted by stars, or be the result of supernova explosions, etc, and are energetic enough to produce secondary particles.

cosmological nucleosynthesis (p 194)
See nucleosynthesis.

cosmological principle (p 182)
The idea that the large scale features of the universe observed by us would also be found by other observers located elsewhere.

critical density (p 185)
Marks the border for the density in cosmological models between two different futures: (i) expansion and (ii) contraction of the model.

degree K (p 10)
A unit of temperature.

demons (p 77)
Imaginary intelligences able to perform tasks which humans cannot accomplish.

deterministic chaos (p 98)
Chaos resulting from situations which do not contain any specific statistical elements. Deterministic equations do not imply predictability.

diffraction (p 129)
The property of waves to move around obstacles. It is exhibited by all forms of waves: acoustical, electrical, optical and quantum mechanical.

disorder (p 114)
A violation of order. Normally statistics is needed to describe it.

effective mass (p 64)
The movement of an electron or hole in a semiconductor is affected by its lattice structure and this can be taken into account to some extent by associating with it a mass different from the normal electron rest mass.

efficiency of an engine (p 27)
Ouput power of a device as a fraction of the input power.

eigenvalue (p 127)
An operator which represents an observable can be written in different mathematically equivalent forms, at least one of which shows explicitly the values which the observable can assume. These are its eigenvalues.

electrolyte (p 154)
Usually an aqueous solution of certain substances which are relatively good conductors of electricity by virtue of free electrons and other charged particles contained in the solution. The passage of a current is normally associated with a chemical reaction.

electron (p 32)
Elementary particle with a negative charge and spin.

element (p 30)
A substance which cannot be decomposed chemically into simpler substances (see table 3.1). Ninety three stable naturally occurring elements exist.

energeticist (p 39)
Scientist who did not believe in the atomic structure of matter, hoping to make do with macroscopic concepts like energy and others taken from thermodynamics. There are not many left.

energy (p 17)
A physical quantity which comes in many different forms such as heat, sound, light, chemical reactions and in magnetic and electrical systems and which can be converted from one form into another.

ensemble (p 140)
This consists of copies of the system of interest which differ form each other and the given system in respect of at least one unspecified aspect of the given system.

entanglement (p 146)
This applies to quantum states of systems having two or more parts. It occurs if it is inadequate to think of a quantum state of such a system merely in terms of quantum states of the separate parts.

entropy (p 70)
A quantity which for an isolated system can stay constant or increase, but cannot decrease. First used in thermodynamics, but now more widely employed.

enzyme (p 116)
A protein which acts as a catalyst (q.v.) for biochemical reactions.

equilibrium state (p 69)
State reached by a system after a sufficient time so that all systematic changes have ceased. Residual fluctuations cannot be eliminated. Normally we mean thermal equilibrium.

escape velocity (p 194)
Minimum velocity which a body must acquire to escape permanently from the gravitational attraction of the parent body.

exclusion principle (p 47)
No two fermions of the same kind can simultaneously occupy the same quantum state.

fermion (p 47)
A particle (not necessarily an elementary one) of an odd number of half integer spins. Only one fermion can occupy a quantum state (see table 3.2). Named after Enrico Fermi.

field (p 55)
A region of space in which a force is exerted on an appropriate body. In a gravitational field a force due to gravity is exerted on any mass. In an electric field an electric force is exerted on an electrically charged body, etc.

fine structure constant (p 217)
A famous constant of approximate value 1/137. It was introduced in connection with the 'splitting' of energy levels of the hydrogen atom by relativistic effects.

fingerprint (p 34)
A term used in this book for the characteristic frequencies of radiation emitted by a material, and which serve to identify it.

flatness problem (p 191)
Why is the average energy density of the universe such that the future of the universe appears to be approximately balanced between a recontraction and an everlasting expansion?

frequency (p 34)
The number of complete oscillations per second of a phenomenon which repeats itself exactly. The lower the frequency of a wave, the longer its wavelength.

galaxy (p 171)
A system of millions or millions of millions of stars, or more, held together by gravity.

game theory (p 229)
A body of knowledge for the analysis of various conflict situations.

gas, ideal (p 11)
A gas in which the interactions among the components are almost negligible.

general theory of relativity (p 158)
A theory which derives gravitational forces from geometrical deformations of a four-dimensional space constituted of time and the three dimensions of space.

geometrical optics (p 127)
A study of light in which its progress is represented as an approximation by one-dimensional lines, straight or curved, but not by waves.

gluon (See table 3.2) (p 53)
Particle which mediates the strong force and binds quarks into elementary particles.

grand unified theory (GUT) (p 58)
The supposition that in the early universe the four forces were unified in a single force. The distances involved were then far smaller than any which have so far been accessible to experiment. The temperatures and energies involved were also very large. Upon expansion the individual forces are believed to develop from this single force. A theoretical description of this situation has not been achieved.

gravitational red shift (p 158)
The downward shift in frequency or energy if a signal or a photon travels upwards in a gravitational field. If it travels downwards it suffers a blue shift.

graviton (p 55)
The quantum of the gravitational field, assumed to be massless and of spin two, but not yet identified experimentally. It mediates the gravitational force and has not yet been identified experimentally.

hadron (p 56)
Baryons and mesons are collectively called hadrons and consist of quarks bound together by gluons (table 3.2). Examples are protons and neutrons.

Hall effect (p 64)
The transverse voltage developed across a current-carrying semiconductor in a magnetic field. Named after E H Hall.

heat capacity (p 196)
The quantity of heat required to raise the temperature of one gram of a material by one degree Centigrade.

heat death (p 73)
The thermal equilibrium state eventually expected in a static or ever-expanding universe model. All life will then have ceased.

hidden variable (p 141)
A variable which is not part of normal quantum theory, yet is supposed to complete the description of a quantum state, thus eliminating the need to use probabilities.

Higgs boson (p 63)
A predicted, so far undiscovered, particle, named after P W Higgs of Edinburgh University, and associated with the origin of mass of all known elementary particles.

holes (p 65)
In a semiconductor refer to electron vacancies. The semiconductor is p-type (for 'positive') if holes dominate over electrons. Otherwise it is n-type (for 'negative').

homogeneous body (p 185)
A body every volume element of which has the same properties.

horizon problem (p 191)
Why is the large-scale universe apparently so homogeneous (q.v.) and isotropic (q.v.)?

Hubble time (p 175)
The reciprocal of the Hubble parameter. The latter gives the recession velocity of a galaxy when it is multiplied by its distance from us.

hypercycle (p 120)
A combination of reaction cycles suggested as important for the origin of self-replicating systems using a form of autocatalysis.

ideal gas (p 11)
A hypothetical gas with only very weak intermolecular forces. It can be approximated by dilute real gases.

impotence, principles of (p 28)
Statements asserting the impossibility of achieving certain things, e.g. violating the second law of thermodynamics or energy conservation.

incompatible (observables) (p 126)
The accuracy in the measurement of pairs of these observables is limited by quantum theory, notably by the uncertainty principle.

incomplete mathematical theories (p 207)
Theories which contain at least one proposition which can neither be proved nor disproved, i.e. which is undecidable.

inconsistent mathematical theories (p 207)
Theories which contain at least one proposition which can be both proved and disproved.

inert gas (p 35)
See Noble gas.

inertial observer (p 155)
One who finds that in his frame of reference the following law holds: a body in motion (and not acted upon by an outside force) will continue with a steady speed and in a straight line. If it is at rest, it will continue at rest. This is also called Newton's first law.

inflation in cosmology (p 191)
A theory suggesting that a short period of very rapid expansion occurred shortly after the Big Bang. This is believed to alleviate the horizon problem and the flatness problem.

information theory (p 147)
Leads to numerical estimates of how the receipt of information can decrease uncertainty. It utilizes analogues of entropy.

instability (of boundary conditions) (p 81)
If a problem leads to certain solutions with a certain boundary condition (q.v.), a completely different solution may be found when the condition is changed to the slightest degree. Such boundary conditions are unstable.

interference (p 131)
The combination of wave trains to produce one or more new waves. Points at which there is no vibration (nodes) are produced by 'destructive interference'.

irreversibility of a process (p 85)
A one-way time evolution which can be turned back in time not simply, but only with considerable outside help. This is not yet fully understood in terms of microscopic physics.

isolated system (p 22)
A system which can exchange neither energy nor matter with its surroundings.

isotope (p 32)
Isotopes of an element differ in the number of neutrons in the nucleus, though the number of protons in the nucleus are the same. The neutral atoms of isotopes of the same element all have the same number of electrons and they therefore have rather similar chemical properties.

isotropic body (p 185)
A body having the same properties in all directions.

kaons (p 76, see table 3.2)
Known carriers of the strong force, which can be neutral or charged.

kaon decay (p 76)
A tiny number of disintegrations of some neutral kaons (see table 3.2 and figure 3.5) were inferred, rather indirectly, in 1964 and directly in 1998, to violate time reversibility. In principle, the puzzle of macroscopic irreversibility could be due to this effect, but this is unlikely.

kinetic energy (p 17)
Energy of motion.

Lamb shift (p 154)
A theoretical and experimental estimate of a tiny gap between two energy levels of the hydrogen atom (and other atoms later). It showed how it was possible to conduct calculations in spite of infinities occurring in the theory. Named after Willis Lamb (b 1913; NL 1955).

large number hypothesis (LNH) (p 219)
Two very large pure numbers occur in physics and are of the same order, we may assume that they should always be equal, since the other pure numbers which occur are so much smaller. This hypothesis was suggested by P A M Dirac (1902–1984).

laser (p 136)
A device which generates intense and well-directed beams of radiation in a narrow frequency range. It is an abbreviation of 'light amplification by stimulated emission of radiation'.

lattice point (p 73)
Crystalline solids (as distinct from amorphous solids, for example) have their atoms arranged on a three-dimensional lattice or grid, with the atoms placed on intersections of the grid lines.

lepton (p 52)
Fermions of spin $\frac{1}{2}h$ that do not carry a colour charge.

laws of thermodynamics (p 23)
See Thermodynamics.

light year (p 171)
The distance travelled by light in vacuum in one year, equal to 9.46 million million kilometres.

locality (p 64, 143)
The notion that events can influence only such other events as are in their immediate vicinity. Alternatively, locality for events implies that they can be connected by messages moving no faster than light.

logistic equation (p 101)
Describes the change with time of population numbers such that one parameter controls whether or not chaos ensues for different initially chosen populations.

macroscopic physics (p 8)
Deals with objects which are human-size or larger.

matrix mechanics (p 136)
See Operator.

Maxwell's demon (p 80)
A demon which creates temperature differences by the operation of trap doors in a system previously at uniform temperature.

meson (p 52)
A particle consisting of a quark and an antiquark (see table 3.2).

messenger particle (p 55)
Their interchange can be regarded as responsible for the forces between objects. See, for example, Graviton.

metastable state (p 100)
A state of a system which can as a result of a small disturbance be converted to another stable state.

microscopic (p 8)
Of molecular or smaller dimension.

microwave background radiation (p 184)
See Cosmic background radiation.

momentum (p 18)
The quantity responsible for the exertion of pressure on a surface. More 'scientifically', it is the quantity whose rate of change is the force acting on the system.

multiverse (p 226)
The whole of physical reality, including and beyond the perceptible 'universe'. This leads to the possibility of many universes, but some argue that the multiverse is identical with the universe.

neutrino (p 59)
An electrically neutral spin ½ fermion of very small mass (see table 3.2). They come in different types.

neutron (p 32)
An uncharged spin ½ fermion whose mass approximates that of a proton.

neutron star (p 174)
A very dense (of the order of a million million million kilograms per cubic metre) star consisting largely of neutrons. For theoretical reasons its total mass cannot greatly exceed the mass of the sun.

Newtonian mechanics (p 55)
Also known as classical mechanics and defined by three main laws. For one of these see inertial observer.

Newton's theory of gravitation (p 55)
Any two masses attract each other by a force proportional to the product of the masses. For point particles the force is also proportional to the inverse square of the distance between them. It explained planetary motion, the tides, etc.

noble gas (p 35)
Gases which were hard to liquefy in the 19th century. 'Inert' and 'rare' are also used, but they are not as suitable as compounds of xenon have been prepared and argon is plentiful.

non-equilibrium steady state (p 100)
A system which changes in a manner which keeps its physical characteristics fixed in time.

non-locality (pp 64, 113)
Absence of locality. It occurs in quantum theory when a particle is affected by conditions not only at one point but also by conditions elsewhere.

nucleosynthesis (p 194)
The fusion of atomic nuclei by nuclear reactions. Cosmological or primordial nucleosynthesis yields the lighter elements like hydrogen, helium and lithium; stellar nucleosynthesis takes place in stars and produces heavier elements ranging from lithium to iron; in supernovae the heavier elements are formed.

nucleus (p 32)
The central massive part of an atom, consisting of neutrons and protons.

observable (p 124)
A physical quantity capable of being measured.

Olbers' paradox (p 187)
The fact that the night sky is dark. Named after Heinrich Olbers who lived at a time when the universe was believed to be static with an infinity of stars. This would be expected to lead to a bright night sky.

omega minus baryon (p 61)
A heavy baryon. See table 3.2 and figure 3.6.

open system (p 22)
A system which can exchange energy and matter with its surroundings.

operator (or matrix) (p 127)
In the context of this book, a mathematical object which can represent a physical observable in the quantum mechanical formalism. A matrix is a rectangular array of numbers with standard rules of manipulation.

order (p 114)
This is exhibited by a system if its physical characteristics are entailed by its specification without the need to appeal to statistics. Example: a lattice is ordered but may have lattice defects whose description requires statistics.

Pascal's wager (p 229)
A kind of bet that you should order your life on the assumption that God exists. You are then supposed to have a better chance of getting to heaven.

periodic table (p 41)
A systematic arrangement of the chemical elements, as in table 3.1.

paramagnet (p 110)
A substance whose atomic magnetic moments can be lined up by an external magnetic field to become a magnet.

Penrose mechanism for extracting energy (p 197)
A proposed means of extracting energy from a rotating black hole. Named after Roger Penrose.

permanent gas (p 35)
See Noble gas.

perpetual motion engines (p 17)
Of the first kind: engines which will work indefinitely, without energy supply, once set into motion.
Of the second kind: engines able to convert heat from the surroundings into mechanical work without loss of energy.

phase space (p 83)
In figure 2.1 a value of the Fahrenheit temperature and a value of the Centigrade temperature determine a point on the graph. If the state of a system depends on more variables (say n), we obtain an n-dimensional phase space for the system, whose state is then represented by a point in this phase space. Its movement in time describes the development of the system.

phase transition (p 107)
This occurs if an abrupt change occurs as a result of either a small parameter change (in theoretical work) or a small change in an external variable (in experimental work).

phonon (p 165)
A quantum of vibrational energy in a solid or elastic medium.

photoelectric effect (p 160)
The ejection of electrons from a metal surface as a result of incident radiation. Associated with the particle aspect of photons (or light).

photon (p 89)
The zero-mass carrier of the electromagnetic force. It can come in all frequencies of the rainbow, and all light (e.g. starlight) consists of photons.

physical optics (p 127)
The study of optical effects talking into account the wave nature of light (i.e. electromagnetic waves).

Planck's constant (p 46)
Relates energy to frequency. Named after Max Planck.

plasma (p 164)
A gas of charged and neutral particles. The interactions among the particles give rise to their interesting 'collective' behaviour.

proper value (p 127)
See Eigenvalue.

proportionality (p 10)
This refers to two quantities such that one increases linearly as the other increases, as in the graph of figure 2.1.

proton (p 32)
A fermion of positive electrical charge found, for example, at the centre of atoms.

quantum (p 9)
The smallest part of a quantity which has its values restricted to integral (or other) multiples of an appropriate unit.

quantum electrodynamics (QED) (p 56)
A quantum field theory which deals with the interactions of electrically charged particles, as well as photons. See box 3.6.

quantum field theory (p 56)
A theory describing particle interactions through the mediation or exchange of messenger bosons.

quantum gravity (p 136)
Attempts to describe gravity in terms of quantum field theory, i.e. by combining quantum theory and relativity.

quantum mechanics (p 140)
The mechanics which has to replace classical mechanics when small distances or elementary constituents of physics are to be described theoretically.

quantum number (p 47)
A quantity needed to characterize the quantum state of a physical system. It usually has an integral or half-integral value.

quantum state (p 45)
The condition of a physical system described by a wavefunction.

quantum theory (p 122)
The universally accepted theory which attributes both wave-like and particle-like characteristics to both matter and radiation, and operates with probabilities derived from wavefunctions. It assigns discrete energy levels to finite systems, and this 'quantization' usually involves Planck's constant.

quark (p 52)
There are six quarks as well as their anti-particles. Each comes in three colours. They make up protons, neutrons, etc, and their electric charge is a fraction of the electron charge. They mediate the strong force.

quasi-particles (p 64)
Particles in such strong interaction that it is no longer satisfactory to treat them as individual units.

rare gas (p 35)
See Noble gas.

ray optics (p 127)
See Geometrical optics.

real gas (p 149)
A gas in which the interactions among the components are taken into account.

red shift (blue shift) (p 171)
The downward shift in frequency of light emitted by a receding source. A shift to higher frequencies occurs (blue shift) for an approaching source. See also gravitational red shift (blue shift).

relativistic speed (this glossary e.g. **cosmic rays**)
Speed close to that of light.

relic radiation (p 184)
See Cosmic background radiation.

resonance (p 97)
If the periodic time (or frequency) of an oscillation of a system and the corresponding quantity of the driving system are very close the resulting oscillation can have a large amplitude and the two systems are said to be in resonance. A strong musical note can shatter a wine glass by this mechanism.

reversibility (p 75)
Applies to phenomena which can consistently occur also in a reverse time direction.

scale factor (p 179)
A function of time which determines how certain cosmological parameters change with the evolution of the universe.

Schrödinger's cat (p 145)
This refers to a hypothetical experiment in which a cat appears to be in a superposition of states some of which make it a live cat, while others make it a dead cat. Named after E Schrödinger.

screening (p 153)
The reduction of a force in space due to intervening bodies.

self-organization (p 108)
This occurs when a large number of small and mutually interacting objects exhibit structure as a result of an external and often small change.

semiconductor (p 65)
A solid which is an insulator at low temperatures by virtue of an energy gap separating occupied from unoccupied electron states. Excitation of electrons into conducting states requires a supply of energy and yields an n-type material. If electron vacancies dominate it is a p-type material.

Shandy, Tristram (p 58)
A character, invented by Laurence Sterne, who, trying to write down his biography, finds that he cannot complete it.

singularity (p 200)
This occurs for the purposes of this book if a variable becomes infinite.

solar cell (p 27)
A semiconductor material capable of furnishing an electric current merely by the influence of radiation, typically from the sun.

space component (p 157, see figure G.2 below.)
The straight line OP is drawn on a grid of lines at right angles to each other, called graph paper, as shown. OA and OB are called its space components. You can reach P from O by going along OA and then along AP, which is the displaced version of OB. If repeated for three space dimensions by starting with the line OQ (instead of OP) you obtain the three space components of OQ, namely OA, OB and OC. Such three-component objects are called vectors. You can have more components and need not confine your attention to ordinary space. An n-component vector exists in an imagined 'n-dimensional space'.

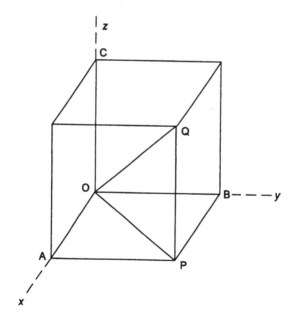

Figure G.2

special theory of relativity (p 157)
Asserts (i) the equivalence of reference systems moving uniformly relative to each other, (ii) the constancy of the velocity of light in vacuum. Simultaneity of events depend on the motion of the observer.

specific gravity (p 35)
The number of times a cubic metre of a material is heavier than an equal volume of water at standard conditions.

spectral lines (p 34)
See Fingerprint.

spin (p 46)
The intrinsic angular momentum of fundamental particles. Its value is always a numerical multiple of Planck's constant.

stable state (p 100)
This is a state to which a system will return spontaneously after an arbitrary small disturbance.

states of matter (p 164)
Conventional division of matter into distinct types such as gases, liquids and solids, together with plasmas, amorphous materials, etc.

statistical mechanics (p 60)
Uses probabilities to relate the properties of normal or larger systems to those of molecules or smaller components.

steady state (p 100)
See Non-equilibrium steady state.

strangeness (p 61)
A kind of quantum number for quarks, see table 3.2.

string theory (p 48)
The speculation that the behaviour of subatomic particles is determined by the arrangement and properties of tiny one-dimensional objects called strings.

strong force (p 56)
The force which keeps nuclei together, governs the interaction between quarks and is transmitted by gluons. It is studied in quantum chromodynamics (QCD).

superatom (p 164)
Several atoms linked by mutual interactions at low temperatures as part of Bose–Einstein condensates.

superconductivity (p 146)
The phenomenon of the loss of electrical resistance by some metals at very low temperatures.

superfluidity (p 146)
The ability of a fluid at very low temperatures to move without friction, and even to climb out of its container.

superposition principle (p 137)
Any number of quantum states of a system may be combined to form a resulting state whose properties are intermediate between the constitutive states. Only one of the states is possible classically.

supersymmetry (SUSY) (p 48)
Transformation rules which are useful in the unification of the physical theory of the distinct forces of nature and of fermions and bosons.

theory of everything (TOE) (p 58)
A single theory, not yet available, of all interactions, including a combination of quantum theory and relativity.

thermal equilibrium (p 161)
This occurs when two bodies, isolated from the surroundings, have reached a definite state.

thermodynamic cycle (p 22)
This occurs when a working fluid is taken through a series of changes which end again with the initial state. The objective is often to transform heat or another form of energy into mechanical or electrical work.

thermodynamic scale of temperature (p 11)
See Absolute scale.

thermodynamics (p 13)
A science developed to discuss the properties of systems of normal size, using quantities like energy, entropy, temperature, pressure, work, etc. The first law (section 2.4) says that energy is conserved in isolated systems; the second law that a *perpetuum mobile* of the second kind (section 2.6) cannot be constructed; the third law (section 2.1) that the absolute zero of temperature cannot be attained.

time dilatation (p 202)
Time passes more slowly for an object which moves fast in an inertial frame than it does for an object fixed in that frame. In particular this holds for a person moving away from a fixed person and then returning. The traveller has aged less as a result.

time machine (p 202)
An imagined non-existent device enabling one to travel into the past.

time's arrow (p 87)
Exhibited by the one-way property of processes like diffusion, egg timers, waterfalls, etc, which need a special effort to be reversed.

tunnel effect in quantum mechanics (p 138)
The circumstance that small enough particles can pass through barriers and walls which are classically impassable.

transmutation (p 31)
The change of one element into another either naturally (as in radioactive decay) or artificially induced (for example, by bombardment).

Turing machine (p 209)
Essentially an early design for a computer and its program. Named after Alan Turing.

twin 'paradox' (p 201)
See Time dilatation.

uncertainty relation (p 124)
This states that certain pairs of physical variables do not have well-defined values simultaneously. The more accurately one is known the more uncertain is the other.

unobservables (p 123)
Potential observables introduced to aid physical theory at a time when they have not (yet?) been observed.

Unruh effect (p 155)
The circumstance that a uniformly accelerated observer in a vacuum close to absolute zero is expected to record a temperature proportional to the acceleration. Named after W G Unruh.

vacuum (p 154)
A region of space in which the only forces and activities are due to the creation and decay of particle–antiparticle pairs. They are called virtual particles as they exist only for tiny periods of time. Field theory recognizes various types of vacuum. If energy is injected a virtual particle may become a 'real' particle.

vector (p 157)
See Space component.

virtual (p 154)
See Vacuum.

wavefunction (p 125)
A function of position and time, which applies to a physical situation in quantum mechanics. It is in general a sum of terms (a 'super-position'), each representing some quantum state, and only one of which describes normally the state after a measurement has been made. This jump is called 'wave function collapse'. The wavefunction furnishes a means of calculating the probabilities of a certain events occurring.

wavelength (p 126)
The distance separating successive crests in a wave. The bigger the wavelength, the smaller the frequency of a wave.

wave mechanics (p 136)
A theory, identical to quantum mechanics, but with an emphasis on wave properties.

wavicle (p 46)
Elementary constituents of matter, emphasizing that they can have both particle and wave properties. The more usual term is 'particle'.

weak force (p 57)
Responsible for the decay of some atoms and for the interconversion of quarks.

zero-point energy (p 162)
Energy which remains in a system even in the limit as the absolute zero of temperature is approached.

zero-point motion (p 162)
The motion, typically oscillations, left in a system even close to the absolute zero of temperature.

References

[1.1]   Eddington A S 1935 *The Nature of the Physical World* (London: Dent) The Gifford Lectures (Original edition 1928)

[1.2]   Horgan J 1996 *The End of Science* (Reading, MA: Addison-Wesley (1997 London: Little Brown)).

[1.3]   Lindley D 1981 *The End of Physics* (1993 New York: Basic Books), Hawking S W 1981 Is the end in sight for theoretical physics *Phys. Bull.* **32** 15

[1.4]   Fukuyama F 1992 *The End of History and the Last Man* (New York: The Free Press)

[1.5]   Maddox J 1998 *What Remains to be Discovered* (London: Macmillan)

[2.1]   Snow C P 1959 *Two Cultures and the Scientific Revolution* The Rede Lecture, Cambridge University Press
1964 *The Two Cultures: A Second Look* (Cambridge: Cambridge University Press)

[2.2]   Soulen R J and Fogl W E in *Physics Today* August 1996 p 36

[2.3]   Feynman R P 1985 QED—the Strange Story of Light and Matter (Princeton University Press) p 6

[2.4]   Mendelssohn K 1973 The World of Walther Nernst (London: Macmillan) p 65

[2.5]   Cropper W H 1987 Walther Nernst and the Last Law *J. Chem. Ed.* **64** 3

[2.6]   Carlip S 1997 Spacetime foam and the cosmological constant *Phys. Rev. Lett.* **79** 4071

[2.7]   Landsberg P T 1998 An unattainability conjecture *Superlattices and Microstructures* **23** 511

[2.8]   Dolby R G A 1996 *Uncertain Knowledge* (Cambridge: University Press)

[2.9]   Larsen E 1953 An American in Europe (New York: Rider & Co) Brown S C 1964 Count Rumford (London: Heinemann)

[2.10]   Bejan A 1988 *Advanced Engineering Thermodynamics* (New York: Wiley) p 408

Angulo-Brown F and Paez-Hernandez R 1993 Endoreversible
thermal cycles with a non-linear heat transfer law *J. App. Phys.* **74**
2216

[2.11]   Wentz F J and Schabel M 1998 Effects of orbital decay on satellite-
derived lower-tropospheric temperature trends *Nature* **394** 661
[2.12]   Landsberg P T 1979 Energy unit for global use *Nature* **178** 502
[2.13]   World Energy Council *Energy for Tomorrow's World* (New York:
St Martin's Press, 1993)
[2.14]   Cole G H A 1996 The provision of global energy *Entropy and
Entropy Generation* ed J S Shiner (Dordrecht: Kluwer) pp 159–173

[3.1]   Weinberg S 1993 *The Discovery of Subatomic Particles* (London:
Penguin)
[3.2]   Penrose R 1989 *The Emperor's New Mind* (Oxford: Oxford
University Press)
[3.3]   Hofstadter D R 1979 *Gödel, Escher, Bach. The Eternal Golden
Braid* (New York: Basic Books)
Dyson F J 1979 *Disturbing the Universe* (New York: Harper and
Row)
[3.4]   Hawking S W 1988 *A Brief History of Time* (Bantam)
[3.5]   Midgley M 1992 *Science as Salvation* (London: Routledge)
[3.6]   Pagels H R 1985 *Perfect Symmetry* (London: Penguin)
[3.7]   Whitehead A N 1946 *Science and the Modern World* (Cambridge:
Cambridge University Press) p 125
[3.8]   Brush S G 1996 The reception of Mendeleev's Periodic Law in
America and Britain *Science in Context* **87** 595
[3.9]   Maxwell J C 1873 Molecules *Nature* **253** 437
[3.10]   Sir James Chadwick 1954, The Rutherford Memorial Lecture, *Proc.
Roy. Soc. A.* **224** 435
[3.11]   Korber H-G 1961 *Aus dem wissenschaftlichen Briefwechsel W
Ostwald*, Vol I (Berlin: Akademie Verlag) pp 22, 118–119
[3.12]   Tod K P 1994 Mach's principle revisited *Gen. Rel. Grav.* **26** 103
[3.13]   Goss Levi B Elements 104–109: soon to be named? *Physics Today*
50 May 1997 p 52
[3.14]   Howorth M 1958 *The Life of Frederick Soddy* (London: New
World) p 83
[3.15]   Abraham Pais 1986 *Inward Bound* (Oxford: Clarendon Press) p 12
[3.16]   The issue of *Synthèse* **119** No 1–2 (1999) is devoted to Boltzmann.
See also his *Lectures on Gas Theory* (Tr. S G Brush, University of
California Press, 1964) p 56
Planck M Verh. d. Deutsch *Phys. Ges.* **2** 202–4
[3.17]   Liboff R L 1984 The correspondence principle revisited *Physics
Today* 8 February 1984 p 50

282 References

[3.18]   Landsberg P T *The Times* (London) February 1974
[3.19]   Close F 1983 *The Cosmic Onion* (London: Heinemann)
[3.20]   Weinberg S 1993 *Dreams of a Final Theory* (London: Hutchinson)
[3.21]   Kane G 1995 *The Particle Garden* (Reading, Mass: Addison-Wesley)
[3.22]   Chown M 1998 Blessed is the Weak *New Scientist* 8 August
[3.23]   Wilczek F Back to basics at ultrahigh temperatures *Physics Today* April 1998 p 11
[3.24]   Miller D J 1996 Beauty stays as charm wilts *Nature* **382** 673
[3.25]   Dennis Normile 1998 Weighing in on neutrino mass *Science* **280** June p 1689
[3.26]   Lakes R 1998 Experimental limits on the photon mass and the cosmic magnetic vector potential *Phys. Rev. Lett.* **80** 1826
[3.27]   Gamow G 1966 *Thirty years that shook physics* (London: Constable and Co. Ltd). Dover edition 1985 (see the Appendix)
[3.28]   Weinberg S 1997 The first elementary particle *Nature* **386** 213

[4.1]   Goldstein M and I F 1993 *The Refrigerator and the Universe* (Cambridge, MA: Harvard University Press)
[4.2]   Landau L D and Lifshitz E M 1958 *Statistical Physics* (London: Pergamon) (and later editions)
[4.3]   Landsberg P T (ed) 1982, 1985 *The Enigma of Time* (Bristol: Adam Hilger) p 119
[4.4]   Landsberg P T and Mann R B 1993 New types of thermodynamics from (1+1)-dimensional black holes *Class. Quantum Grav.* **10** 2373
[4.5]   Wigner E P 1980 *Some Strangeness in the Proportion* (ed H Woolf (Reading, MA: Addison Wesley) p 467
[4.6]   Landsberg P T (ed) *Proc. Int. Conf. on Thermodynamics (Cardiff, 1970)* (London: Butterworths). (Also in 1970 *Pure Appl. Chem.* **22** p 543)
[4.7]   Leff H D and Rex A F 1990 *Maxwell's demon* (Bristol: Hilger) von Bayer H C 1998 *Maxwell's Demon* (New York: Random House)
[4.8]   Denbigh K G 1981 *Three Concepts of Time* (Berlin: Springer)
[4.9]   Shallis M 1983 *On Time* (Penguin Books)
[4.10]   Schumacher B W 1994 Demonic heat engines *Physical Origins of Time Asymmetry* ed J J Halliwell, J Perez-Mercader and W H Zurek (Cambridge: Cambridge University Press) p 90
[4.11]   Aharony A 1971 Time reversal symmetry violation and the H-theorem *Phys. Lett.* **37A** 45
[4.12]   Zeh H D 1990 in *Complexity, Entropy and the Physics of Information* ed W H Zurek (Reading, MA: Addison Wesley)
[4.13]   Denbigh K G and J S 1985 *Entropy in Relation to Incomplete Knowledge* (Cambridge University Press) p 44ff

[4.14] Gibbs J W 1902 *Elementary Principles of Statistical Mechanics* (New Haven, CT: Yale University Press; reprinted by New York: Dover) p 145

[4.15] Bohm D 1980 *Wholeness and the Implicate order* (London: Routledge and Kegan Paul) p 179

[4.16] Zhang S, Meier B H and Ernst R R 1992 Polarization echoes in NMR *Phys. Rev. Lett.* **69** 2149

[4.17] Sklar L 1993 *Physics and Chance* (Cambridge University Press) p 417

[4.18] Karakostas V 1996 On the Brussels school's arrow of time in quantum theory *Philosophy of Science* **63** 374

[4.19] Speziale P (ed) 1972 *Albert Einstein–Michele Besso Correspondence* (Paris: Hermann) p 538

[4.20] Quentin Smith 1993 *Language and Time* (New York: Oxford University Press)

[4.21] Huw Price 1996 *Time's Arrow and Archimedes' Point* (New York: Oxford University Press)

[4.22] Halliwell J J, Perez-Mercader J and Zurek W H (eds) 1994 *Physical Origins of Time Asymmetry* (Cambridge: Cambridge University Press)

[4.23] Tanner T 1970 The American novelist as an entropologist *London Magazine* **1** 5

[4.24] Harrison M J The ash circus *New Worlds* (April 1969) p 18

[4.25] Greenland C 1983 *The Entropy Exhibition* (London: Routledge and Kegan Paul)

[4.26] Arnheim R 1971 *Entropy and Art* (University of California Press) Malina F J (ed) 1979 *Visual Art, Mathematics and Computers* (Oxford: Pergamon) pp 27–42

[4.27] Birkhoff G D 1933 *Aesthetic Measure* (Cambridge, MA: Harvard University Press)

[4.28] Georgescu-Roegen N 1971 *The Entropy Law and the Economic Process* (Cambridge, MA: Harvard University Press). Reviewed by M J Green in *The Economic Journal* June 1973 p 551

[4.29] Young J T 1991 Is the entropy law relevant to the economics of natural resource scarcity? *J. Environmental Economics and Management* **21** 169. Comment by K N Townsend in vol 23 p 96

[4.30] Zemach E M 1968 Many times *Analysis* **28** 145

[4.31] Dorling J 1970 The dimensionality of time *Am. J. Phys.* **38** 539

[4.32] Borges J L 1970 The garden of forking paths. Reprinted in *The Traps of Time* ed M Moorcock (Harmondsworth: Penguin) p 156

[5.1] Miles J 1984 Strange attractors in fluid dynamics *Adv. App. Mech.* **24** 189

[5.2] Lighthill J 1986 The recently recognised failure of predictability in Newtonian mechanics *Proc. Soc. A.* **407** 35

[5.3] Thompson J M T 1992 Global unpredictability in nonlinear dynamics: capture, dispersal and the indeterminate bifurcations *Physica D* **58** 260

[5.4] Tritton D J 1986 Ordered and chaotic motion of a forced spherical pendulum *Eur. J. Phys.* **7** 162

[5.5] Hilborn R C 1994 *Chaos and Nonlinear Dynamics* (New York: Oxford University Press) p 40

[5.6] Schurz G 1996 Kinds of unpredictability in deterministic systems; Noyes H P 1996 Decoherence, determinism and chaos revisited. Both papers are in *Law and Prediction in the Light of Chaos Research* ed P Weingartner and G Schurz (Berlin: Springer) p 123 and p 153

[5.7] Douady S and Couder Y 1992 Phyllotaxis as a physical self-organised growth process *Phys. Rev. Lett.* **68** 2098

[5.8] Locke D 1987 Numerical aspects of Bartok's string quartets *Musical Times* June p 322

[5.9a] Markus M and Hess B 1990 Isotropic cellular automaton for modelling excitable media *Nature* **347** 56

[5.9b] Markus M, Müller S C, Plesser T and Hess B 1987 On the recognition of order and disorder *Biol. Cybernetics* **57** 187

[5.10] Scholl E 1987 *Nonequilibrium Phase Transitions in Semiconductors* (Berlin: Springer)

[5.11] Nikolic K 1992 Steady states in a cooled p-Ge photoconductor via the Landsberg–Scholl–Shukla model *Solid-State Electron.* **35** 671

[5.12] Aoki I 1995 Entropy production in living systems: from organisms to ecosystems *Thermochim. Acta* **250** 359

[5.13] Schrödinger E 1944 *What is Life?* (Cambridge: Cambridge University Press). For criticisms, not accepted by all, see Perutz M 1987 *Schrödinger, Centenary celebration of a polymath* ed C W Kilmister (Cambridge: University Press) p 234

[5.14] Prigogine I 1965 *Non-Equilibrium Thermodynamcis, Variational Techniques and Stability* (Chicago, IL: Chicago University Press)

[5.15] Nicolis G and Prigogine I 1977 *Self-Organisation in Nonequilibrium Systems* (New York: Wiley)

[5.16] Nicolis G and Prigogine I 1989 *Exploring Complexity* (New York: Freeman)

[5.17] Haken H 1983 *Synergetics, An Introduction* (Berlin: Springer, 3rd edn)

[5.18] Haken H 1989 Synergetics: An Overview *Rep. Prog. Phys.* **52** 515

[5.19] Mandelbrot B 1977 *The Fractal Geometry of Nature* (New York: Freeman)

[5.20] Landsberg P T 1994 Self-organisation, entropy and order *On Self-Organisation* ed R K Mishra, D Maass and E Zwierlein (Berlin: Springer) p 157

[5.21] Tipler F J 1992 The ultimate fate of life in universes which undergo inflation *Phys. Lett.* B **286** 36

[5.22] Shiner J S, Davison M and Landsberg P T 1999 Simple measure of complexity *Phys. Rev.* E **59** 1459

[5.23] Ebeling W, Engel H and Herzel H 1990 *Selbstorganisation in der Zeit* (Berlin: Akademie Verlag)

[5.24] Thaxton C B, Bradley W L and Olsen R L 1984 *The Mystery of Life's Origin* (New York: Philosophical Library)

[5.25] Wigner E P 1961 The probability of the existence of a self-reproducing unit *The Logic of Personal Knowledge. Essays presented to Michael Polanyi* (London: Routledge and Kegan Paul) Reprinted in Wigner E P 1967 *Symmetries and Reflections* (Bloomington, IN: Indiana University Press) p 200

[5.26] Landsberg P T 1964 Does quantum mechanics exclude life? *Nature* **203** 928
Wigner E P and Landsberg P T 1965 Does quantum mechanics exclude life? *Nature* **205** 1306

[5.27] Levi-Montalcini R 1988 *In Praise of Imperfection; my Life and Work* transl L Attardi (New York: Basic Books Inc.)
Baez J C 1991 Is life improbable? *Foundations of Physics* **19** 91

[5.28] Miller S and Orgel L 1974 *The Origins of Life on the Earth* (Englewood Cliffs, NJ: Prentice Hall)

[5.29] Arkhipov A V 1996 New arguments for panspermia *Observatory* **116** 396

[5.30] Crick F 1981 *Life Itself* (New York: Simon and Schuster)

[5.31] Webber B H and Depew 1996 Natural selection and self-organisation *Biology and Physics* **11** 33

[5.32] Eigen M 1983 Self-replication and molecular evolution *Evolution from Molecules to Man* ed D S Bendall (Cambridge: Cambridge University Press)

[5.33] Campbell N A 1996 *Biology* (Menlo Park, NY: Benjamin/Cumming, 4th edn) Chapter 24: Early earth and the origin of life

[5.34] Lifson S 1987 Chemical selection, diversity, teleonomy and the second law of thermodynamics. Reflections on Eigen's theory of self-organization of matter *Biophys. Chem.* **26** 303

[5.35] Landsberg P T 1972 Time in statistical physics and special relativity *The Study of Time* ed J T Fraser and G H Müller (Berlin: Springer) p 59. Reprinted 1970 *Studium Generale* **23** 1108

[5.36] Born M 1949 *The Natural Philosophy of Cause and Chance* (Oxford: Clarendon) The Waynflete Lectures

[5.37]   Lotka A J 1925 *The Elements of Physical Biology* (Baltimore: Williams and Wilkins; reissued as *Elements of Mathematical Biology* by Dover in 1956)

[6.1]   Mermin D 1997 Sociologists, scientist pick at threads of arguments about science *Physics Today* January p 94

[6.2]   Russell B 1927; 1992 *The Analysis of Matter* (London: Routledge) p 235. Also other editions

[6.3]   Home D and Sengupta S 1983 Heisenberg's gedanken experiment revisited *Am. J. Phys.* **51** 567

[6.4]   Goldstein H 1980 *Classical Mechanics*, 2nd edn (Reading MA: Addison Wesley; 1st edn 1950), Section 10.8

[6.5]   Fishbone P M, Gasiorowicz S and Thornton S T 1996 *Physics* (Englewood Cliffs, NY: Prentice Hall)

[6.6]   de Broglie L and Davisson C J 1965 *Nobel Lectures, Physics 1922–1941* (Amsterdam: Elsevier) p 247 and p 388

[6.7]   Fujikawa F and Ono Y A (eds) 1996 *Quantum Coherence and Decoherence* (Amsterdam: Elsevier)

[6.8]   Born M 1954 The statistical interpretation of quantum mechanics *Science* **122** 675; 1964 *Nobel Lectures, Physics 1942–1962* (Amsterdam: Elsevier)

[6.9]   Whitaker A 1996 *Einstein, Bohr and the Quantum Dilemma* (Cambridge: Cambridge University Press). Author's usual initials: M A B

[6.10]   Einstein A, Podolsky B and Rosen N 1935 Can quantum mechanical description be considered complete? *Phys. Rev.* **47** 777, the so-called EPR paper; see also Einstein A 1936 Physik und Realität *J. Franklin Inst.* **221** 313

[6.11]   Goldstein S 1998 Quantum theory without observers *Physics Today* March p 42 and April p 38

[6.12]   Gottfried K 1991 Does quantum mechanics carry the seeds of its own destruction? *Phys. World* October p 34

[6.13]   Anderson P W 1998 book review in *Phys. World* August p 56

[6.14]   Ekert A 1995 Pet theories of quantum mechnics *Phys. World* December

[6.15]   Lahti P and Mittelstaedt P (eds) 1985 *Symp. on the Foundations of Modern Physics: 50 years of the Einstein–Podolsky–Rosen Gedankenexperiment* (Singapore: World Scientific) Greenberger D and Zeilinger A 1995 Quantum theory: Still crazy after all these years *Phys. World* September p 33

[6.16]   Mermin N D 1993 Hidden variables and the two theorems of John Bell *Rev. Mod. Phys.* **65** 803

[6.17]   Peres A 1993 *Quantum Theory: Concepts and Methods* (Dordrecht: Kluwer)

[6.18] Lindley D 1996 *Where has all the Weirdness gone?* (Harper Collins) p 198

[6.19] Zurek W H 1997 Probing origins of the classical *Phys. World* January p 27, and references cited there. See also letters in *Physics Today* February 1999 pp 11–15, 89–92

[6.20] Slosser J J and Meystre P 1997 Resource Letter CQO-1: Coherence in Quantum Optics *Am. J. Phys.* **65** 275
See also Vedral V and Plenio M 1998 Basics of Quantum Computation *Prog. in Quantum Electronics* **22** 1

[6.21] Wick D 1995 *The Infamous Boundary: Seven Decades of Controversy in Quantum Physics* (Basel: Birkhäuser)

[6.22] Rae A 1986 *Quantum Physics: Illusion or Reality?* (Cambridge: Cambridge University Press) p 61

[6.23] Mirman R 1995 *Group Theoretical Foundations of Quantum Mechanics* (Commack, NY: Nova Science Publishers) p 165

[6.24] van Kampen N G 1991 Mystery of quantum measurement *Phys. World* **4** 16

[6.25] Baltes H P 1976 Planck's radiation law for finite cavities and related problems *Infrared Phys.* **16** 1

[6.26] Katz J and Okuta Y 1995 Why cavities? Containers in black-hole thermodynamics *Class. Quantum Grav.* **12** 2275, and references given there

[6.27] Milonni P W and Shih M-L 1992 Casimir forces *Contemporary Physics* **33** 313
Lamoreaux S K 1997 Casimir force measured *Phys. Rev. Lett.* **78** 5

[6.28] Cole D C and Puthoff H E 1993 Extracting energy and heat from the vacuum *Phys. Rev. E.* **48** 1562

[6.29] Pais A 1986 *Inward Bound* (Oxford: Oxford University Press) p 310

[6.30] McDonald K T 1996 Can an electron *be* at rest? *Am. J. Phys.* **64** 1098

[6.31] Bohm D 1957 *Causality and Chance in Modern Physics* (London: Routledge and Kegan Paul) p 1

[6.32] Unruh W G 1976 Notes on black-hole evaporation *Phys. Rev.* **D14** 870
Davies P C W 1975 Scalar particle production in Schwarzschild and Rindler metrics *J. Phys. A: Math. Gen.* **8** 609

[6.33] See for example Jaekel M-T and Reynaud S 1997 Movement and fluctuations of the vacuum *Rep. Prog. Phys.* **60** 863

[6.34] *Energy Statistics of OECD Countries 1993–4* (OECD: Paris, 1996) p 273

[6.35] Pound R V and Rebka G A 1960 Apparent weight of photons *Phys. Rev. Lett.* **4** 337

[6.36] Clark R W 1971 *Einstein. The Life and Times* (New York: Avon Books)

[6.37]   Vessot R F C *et al* 1980 Test of relativistic gravitation with space-borne hydrogen maser *Phys. Rev. Lett.* **45** 2081

[6.38]   Davis K B, Mewes M-O, Andrews M R, van Druten N J, Durfee D S, Kurn K M and Ketterle W 1995 Bose Einstein condensation *Phys. Rev. Lett.* **75** 3969
Griffin A, Snoke D W and Stringari S (eds) 1995 *Bose–Einstein Condensation* (Cambridge: Cambridge University Press)

[6.39]   Burnett K 1997 The amazing atom laser *Nature* **385** 482

[6.40]   Tinkham M 1996 *Introduction to Superconductivity* (New York: McGraw-Hill)

[7.1]   Eve A S 1939 *Rutherford* (New York: Macmillan) p 107

[7.2]   Burchfield J D 1975 *Lord Kelvin and the Age of the Earth* (London: Macmillan)

[7.3]   Bolte M and Hogan C J 1995 Conflict over the age of the earth *Nature* **376** 399

[7.4]   Bahcall J N Redshifts as distance indicators Fig. III, in Field G B, Arp H and Bahcall J N 1973 *The Redshift Controversy* (Reading, MA: Benjamin)

[7.5]   Speziale P (ed) 1972 *Albert Einstein—Michele Besso Correspondence* (Paris: Hermann) p 453. For a simple exposition, see Home D and Gribbin J What is Light? *New Scientist* 2 November 1991 p 30

[7.6]   Lamb Jr W E 1995 Anti-photon *Appl. Phys. B* **60** 77

[7.7]   Avnir D, Biham O, Lidar D and Malcai O 1998 Is the geometry of nature fractal? *Science* **279** 39 and correspondence p 783ff

[7.8]   Landsberg P T and Evans D A 1977 *Mathematical Cosmology: An Introduction* (Oxford: Oxford University Press) p 34

[7.9]   Thornhill of London NW1 in *Listener* December 1950 p 798

[7.10]   Kragh H 1996 *Cosmology and Controversy* (Princeton, NJ: Princeton University Press)

[7.11]   Hu W, Sugiyama N and Silk J 1997 The physics of microwave background anisotropies *Nature* **386** 37

[7.12]   Longair M S 1993 Modern cosmology—a critical assessment *Q. J. R. Astron. Soc.* **34** 157

[7.13]   Harrison E R 1984 The dark night sky riddle *Science* **226** 941
Maddox J 1991 Olbers' paradox has more to teach *Nature* **349** 363

[7.14]   Tipler F J 1988 Johann Mädler's resolution of Olbers' paradox *Q. J. R. Astron. Soc.* **29** 313

[7.15]   Landsberg P T (ed) 1978 *Int. Conf. on Thermodynamics* (Cardiff) p 543. Also in *Pure Appl. Chem.* **22** No. 3/4

[7.16]   Hawking S W 1988 *A Brief History of Time* (New York: Bantam) p 150

[7.17] Landsberg P T, Piggott K D and Thomas K S 1992 Many-cycle effects in irreversible oscillating universe models *Astro. Lett. Commun.* **28** 235

[7.18] Watson M G 1986 The most reliable black hole yet *Nature* **321** 16

[7.19] White N E and van Paradijs J 1996 The galactic distribution of black hole candidates in low-mass X-ray binary systems *Astrophys. J.* **473** L25

[7.20] Rees M 1997 *Before the Beginning* (London: Simon and Schuster)

[7.21] Rowan-Robinson M 1996 On the wilder shores of cosmology *Nature* **379** 309. A book review

[7.22] Halliwell D, Perez-Mercader J and Zurek W H (eds) 1994 *Physical origins of time assymetry* (Cambridge: Cambridge University Press) Part V

[7.23] Bahcall J N and Ostriker J P 1997 *Unsolved Problems in Astrophysics* (Princeton: Princeton University Press)

[7.24] Brandenberger R, Mukhanov V and Sornborger A 1993 *Phys. Rev. D* **48** 1629

[7.25] Earman J 1995 *Bangs, Crunches, Whimpers, and Shrieks* (Oxford: Oxford University Press)

[7.26] 1924 *Observations by Max Beerbohm* (London: William Heinemann)

[7.27] Thorne K S 1994 *Black Holes and Time Warps: Einstein's Outrageous Legacy* (New York: Norton)

[7.28] Hawking S W 1992 The chronology protection conjecture *Phys. Rev. D* **46** 603

[7.29] Adams F C and Laughlin G 1997 A dying universe: the long-time fate and evolution of astrophysical objects *Rev. Mod. Phys.* **69** 337

[7.30] Peebles P J E, Seager S and Hu W 2000 Delayed recombination *Ap. J.* **539** L1

[8.1] Smullyan R M 1992 *Gödel's incompleteness theorems* (Oxford: Oxford University Press)

[8.2] Shanker S G (Ed.) 1988 *Gödel's theorem in Focus* (London: Croom Helm)

[8.3] Casti J L 1990 *Beyond Belief. Randomness, Prediction and Explanation in Science* ed J L Casti and A Karlquist (Baton Rouge, LA: CRC Press)

[8.4] Rucker R 1988 *Mind Tools* (London: Penguin)

[8.5] Landsberg P T 1954 Paradoxes in N-valued logics *Analysis* **15** 14

[8.6] Penrose R 1994 *Shadows of the Mind* (Oxford: Oxford University Press)

[8.7] Andrew Hodges in Herken R (ed) 1988 *The Universal Turing Machine. A Half-Century Survey* (Oxford: Oxford University Press)

[8.8] Chaitin G J 1988 Randomness in arithmetic *Sci. Am.* July p 80

[8.9] Smith Q 1987 Infinity and the past *Phil. Sci.* **54** 63

[8.10]   Moore A W 1990 *The Infinite* (London: Routledge)

[8.11]   Weinberg S 1993 *Dreams of a Final Theory* (London: Hutchinson Radius)

[8.12]   Davies P C W and Brown J R (eds) 1986. Conversation between D Deutsch and P C W Davies *The Ghost in the Atom* (Cambridge: Cambridge University Press)

[8.13]   Eddington A S 1946 *Fundamental Theory* (Cambridge: Cambridge University Press)

[8.14]   Landsberg P T 1996 Irreversibility and Times's Arrow *Dialectica* **50** 247

[8.15]   Gillies G T 1997 The Newtonian gravitational constant: recent measurements and related studies *Rep. Prog. Phys.* **60** 151

[8.16]   Weinberg S 1983 Overview of theoretical prospects for understanding the values of the fundamental constants *Phil. Trans. R. Soc. A* **310** 249

[8.17]   Eddington A S 1939 *The Philosophy of Physical Science* The Tarner Lectures (Cambridge: Cambridge University Press) p 170

[8.18]   Landsberg P T and Bishop N T 1975 A cosmological deduction of the order of magnitude of an elementary particle mass and of the cosmological coincidences *Phys. Lett. A* **53** 109

[8.19]   Chandrasekhar S 1931 The cosmological constants *Nature* **139** 757

[8.20]   Stewart J Q 1931 Nebular red shift and universal constants *Phys. Rev.* **38** 2071

[8.21]   Palacios J 1964 *Dimensional Analysis* (London: Macmillan) (I am indebted to Professor Manuel Castans of Madrid for this reference.)

[8.22]   Petley B W 1985 *The Fundamental Physical Constants and the Frontier of Measurement* (Bristol: Hilger)

[8.23]   Kilmister C W 1994 *Eddington's Search for a Fundamental Theory* (Cambridge: Cambridge University Press)

[8.24]   Landsberg P T 1984 Mass scales and cosmological coincidences *Ann. Phys., Lpz.* **41** 88

[8.25]   Barrow J D and Tipler F J 1986 *The Anthropic Cosmological Principle* (Oxford: Clarendon)

[8.26]   Bertola E and Curi U (Eds.) 1993 *The Anthropic Principle* (Cambridge: Cambridge University Press)

[8.27]   Gribbin J and Rees M 1990 Cosmic coincidences *New Scientist* 13 January p 51

[8.28]   Harrison E R 1995 The natural selection of universes containing intelligent life *Q. J. R. Astron. Soc.* **36** 193

[8.29]   Chown M 1998 Anything goes *New Scientist* 6 June p 28

[8.30]   Gott III J R 1993 Implications of the Copernican principle for our future prospects *Nature* **363** 315

[8.31]   Gott III J R 1997 Tories out on cue *Nature* **287** 842

[8.32]   Landsberg P T, Dewynne J N and Please C P 1993 *Nature* **365** 384

[9.1] Smith Q 1994 Stephen Hawking's cosmology and theism *Analysis* **54** 236

[9.2] Haugaard W P 1968 *Elizabeth and the English Reformation* (Cambridge: Cambridge University Press) p 25

[9.3] Jeffrey R C 1965 *The Logic of Decision* (New York: McGraw Hill) p 12

[9.4] Landsberg P T 1971 Gambling on God *Mind* **80** 100

[9.5] Larsen E J and Witham L 1998 Leading scientists still reject God *Nature* **394** 313

[9.6] Schrödinger E 1956 *Mind and Matter* (Cambridge: Cambridge University Press). Lecture 5

[9.7] Landsberg P T 1991 From entropy to God? in *Thermodynamics: History and Philosophy; Facts, Trends and Debates* ed L Rapolyi (Singapore: World Scientific)

[9.8] Hughes G J 1995 *The Nature of God* (London: Routledge)
Torrance T F 1981 *Divine and Contingent Order* (Oxford: Oxford University Press)

[9.9] Boltzmann L 1964 *Lectures on Gas Theory* (Berkeley: University of California Press), transl. S G Brush. (The original volume was published in 1898.)

[9.10] Bondi H 1952 *Cosmology* (Cambridge: Cambridge University Press)
Gamow G 1952 *The Creation of the Universe* (New York: Viking Press)

[9.11] Bludman S A 1984 Thermodynamics and the end of a closed universe *Nature* **308** 319

[9.12] Tipler F J 1986 General relativity and the eternal return *Essays in General Relativity* (New York: Academic Press)

[9.13] Israelit M and Rosen N 1989 A singularity-free cosmological model in general relativity *Astrophys. J.* **342** 627
Kardashev N S 1990 Optimistic cosmological model *Mon. Not. R. Astron. Soc.* **243** 252

[9.14] Eddington A S 1935 *New Pathways in Science* The Messenger Lectures (Cambridge: Cambridge University Press) p 59

[9.15] Whittaker E T 1942 *The Beginning and End of the World* (Oxford: Oxford University Press) pp 39–40, 63

[9.16] Hawking S W 1988 *A Brief History of Time* (Toronto: Bantam) pp 136, 140, 141

[9.17] Russell R J, Stoeger W R S J and Coyne G V S J (eds) 1988 *Physics, Philosophy and Theology* (Vatican City State: Vatican Observatory)

[9.18] Penrose R 1981 Time-asymmetry and quantum gravity *Quantum Gravity II* ed C J Isham, R Penrose and D Sciama (Oxford: Clarendon)

292 References

[9.19]   Tryon F P 1973 Is the universe a vacuum fluctuation? *Nature* **246** 396
         Jou D and Pavon D 1989 Thermodynamics and Cosmology
         Foundations of Big Bang Cosmology ed W Meyerstein (Singapore:
         World Scientific)
[9.20]   Page D N 1990 Is the universe infinitely old? *Class. Quantum
         Gravity* **7** 1841
         Craig W L and Smith Q 1993 *Theism, Atheism, and Big Bang
         Cosmology* (Oxford: Oxford University Press)
         Deltete R J and Guy R A 1996 Emerging from imaginary time
         Synthese **108** 185
[9.21]   Zuckerman B and Hart M H (eds) 1995 *Extraterrestrials: Where Are
         They?* (Cambridge: University Press)
         Heidmann J 1995 *Extraterrestrial Intelligence* (Cambridge:
         University Press)
[9.22]   Montefiore H 1985 *The Probability of God* (London: SCM)
[9.23]   Davies P C W 1983 *God and the New Physics* (New York: Simon and
         Schuster)
[9.24]   Ayer A J *Language, Truth and Logic* (London: Gollancz) p 121
[9.25]   Landsberg P T 1981 The search for completeness *Nature and System*
         3 236

[10.1]   Landau I 1995 The paradox of the end *Philosophy* **70** 555
[10.2]   Terner J and Pew W L 1978 *The Courage to be Imperfect; The Life
         and Work of Rudolf Dreikurs* (New York: Hawthorn). (I am
         indebted to the late Dr Joseph Meiers for drawing my attention to
         this book.)
[10.3]   Myers D G and Diener E 1995 Who is happy? *Psychol. Sci.* **6** 10
[10.4]   Rescher N 1984 *The Limits of Science* (Berkeley: University of
         California Press)
[10.5]   Trigg R 1993 *Rationality and Science: Can science explain
         everything?* (Oxford: Blackwell)
[10.6]   Cornwell J (ed) 1995 *Nature's Imagination: The Frontiers of
         Scientific Vision* (Oxford: Oxford University Press)
[10.7]   Bondi H 1977 The lure of completeness *Encyclopedia of Ignorance*
         ed R Duncan and M Weston-Smith (Oxford: Pergamon)
         Popper K R 1974 Scientific reduction and the essential
         incompleteness of all science *Studies in the Philosophy of Biology* ed
         F Ayala and T Dobzhansky (Berkeley: University of California
         Press)
[10.8]   Barrow J D 1998 *Impossibility; the Limits of Science and the Science
         of Limits* (Oxford: Oxford University Press)
         Kaku M 1997 *Visions: How Science will Revolutionise the Twenty-
         First Century* (Oxford: Oxford University Press)

[10.9] Wayne Smith J 1989 The logical limits of science *Epistemologia* **12** 153

[10.10] Einstein A 1933 *Preface to Max Planck, Where is Science going?* (London: George Allen and Unwin)

[10.11] Midgley M 1992 *Science as Salvation; a Modern Myth and its Meaning* (London: Routledge)
Midgley M 1992 Can science save its soul? *New Scientist* 1 August

[10.12] Atkins P 1992 Will science ever fail? *New Scientist* 8 August

[10.13] Aydin Z Z and Yilmazer A U 1988 *Nuovo Cimento A* **99** 85

[10.14] Li F-b 1990 *Helvetica Physica Acta* **63** 922

[10.15] *The Times* 13 October 1998

[10.16] Traweek S 1988 *Beamtimes and Lifetimes: the World of High-Energy Physicists* (Cambridge, MA: Harvard University Press) p 119
Hellman H 1998 *Great Feuds in Science* (New York: Wiley)

[10.17] Sokal A and Bricmont J 1998 *Intellectual Impostures* (London: Profile)
Dawkins R 1998 Postmodernism disrobed *Nature* **394** 141

[10.18] Merton R 1977 *The Sociology of Science* (Chicago, IL: University of Chicago Press)

[10.19] Fuller S 1993 *Philosophy of Science and its Discontents* (New York: Guildford)

[10.20] Cole J R and Cole S 1972 The Ortega Hypothesis *Science* **178** 368

[10.21] Hull D L, Tessner P D and Diamond A M 1978 Planck's Principle *Science* **202** 717

Name Index

Index

Printed and bound by CPI Group (UK) Ltd, Croydon, CR0 4YY

22/10/2024

01777625-0002